Grade 2-3

Summer Activity Playground

12 weeks of Summer Activities:

- Math
- ELA
- Science
- Reading
- Social Studies

Brain Hunter Prep is a division of ArgoPrep dedicated to providing high-quality workbooks for K-8th grade students. We have been awarded multiple awards for our curriculum, books and/or online program. Here are a few of our awards!

Our goal is to make your life easier, so let us know how we can help you by e-mailing us at: info@argoprep.com.

ISBN: 9781951048181
Published by Brain Hunter Prep.

Aknowlegments:
Icons made by Freepik, Creaticca Creative Agency, Pixel perfect , Pixel Buddha, Smashicons, Twitter , Good Ware, Smalllikeart, Nikita Golubev, monkik, DinosoftLabs, Icon Pond from www.flaticon.com

BACK to SCHOOL

- ArgoPrep is a recipient of the prestigious **Mom's Choice Award**.
- ArgoPrep also received the 2019 **Seal of Approval** from Homeschool.com for our award-winning workbooks.
- ArgoPrep was awarded the 2019 **National Parenting Products Award**, **Gold Medal Parent's Choice Award** and a **Brain Child Award**

Want an amazing offer from ArgoPrep?

7 DAY ACCESS

to our online premium content at **www.argoprep.com**

Online premium content includes practice quizzes and drills with video explanations and an automatic grading system.

Chat with us live at **www.argoprep.com** for this exclusive offer.

Summer Activity Playground Series

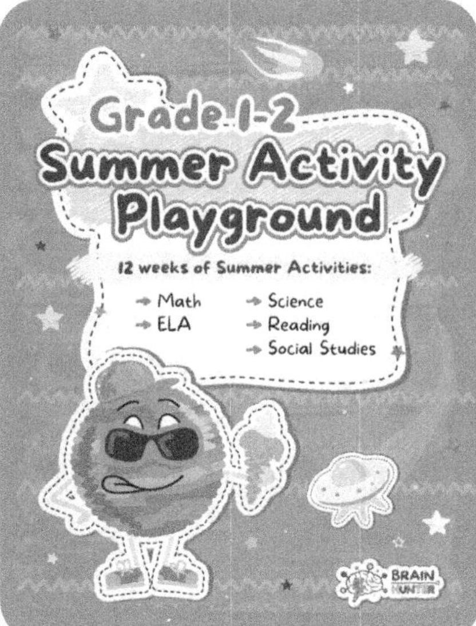

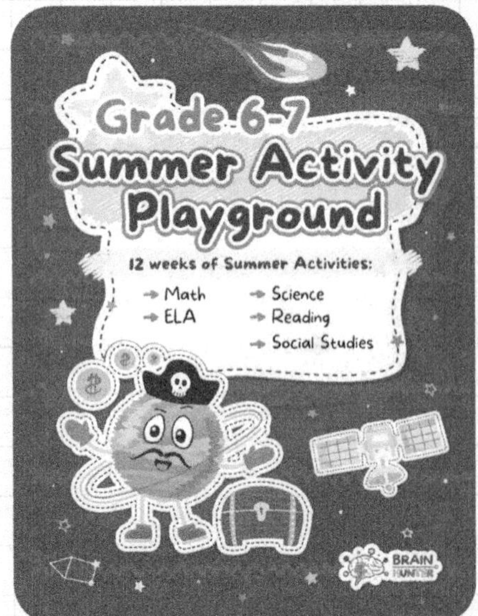

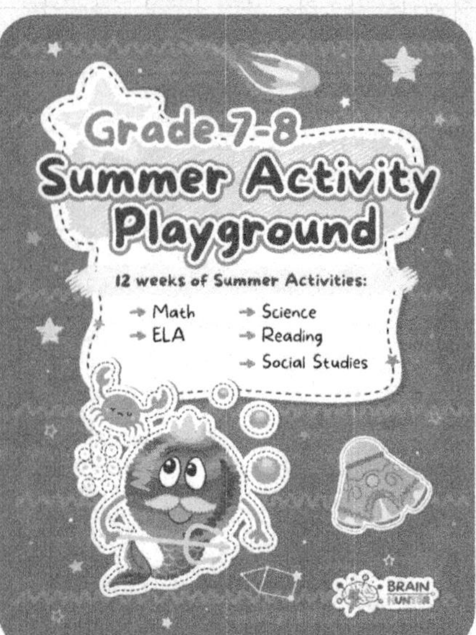

How to use this book?

Welcome to **Summer Activity Playground by Brain Hunter Prep**!

This workbook is specifically designed to prepare students over the summer to get ready for **Grade 3**. Our workbook is divided into twelve weeks so students can complete the entire workbook over the summer.

Our workbooks have been carefully designed and crafted by licensed teachers to give students an incredible learning experience. Students will be able to practice mathematics, english activities, science experiments, social studies, and fitness activities. Give your child the education they deserve!

Summer list to read

We strongly encourage students to read several books throughout the summer. Below you will find a recommended summer reading list that we have compiled for students entering into Grade 3. You can see this list at: www.argoprep.com/**summerlist**

Author: Louis Sachar
Title: Marvin Redpost

Author: Megan McDonald
Title: Judy Moody

Author: Cynthia Rylant
Title: Henry and Mudge

Author: Barbara Robinson
Title: The Best School Year Ever

Author: E. B. White
Title: Charlotte's Web

Author: Mary Pope Osborne
Title: Summer of the Sea Serpent

Author: Liesl Shurtliff
Title: Jack: The True Story of Jack and the Beanstalk

Author: Katherine Applegate
Title: The One and Only Ivan

Author: Beverly Cleary
Title: The Mouse and the Motorcycle

Author: Patricia MacLachlan
Title: Sarah, Plain and Tall

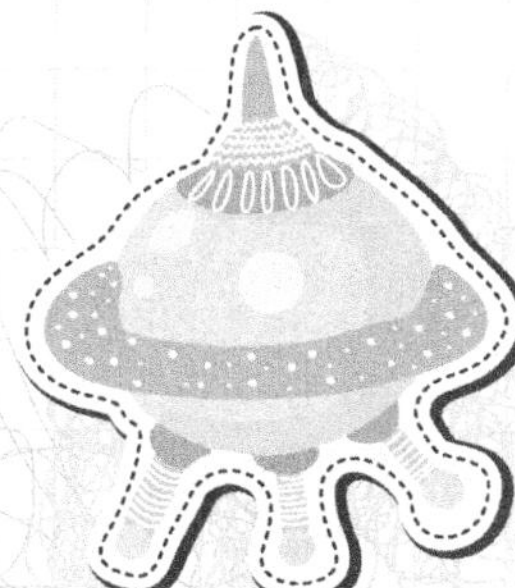

OTHER BOOKS BY ARGOPREP

Here are some other test prep workbooks by ArgoPrep you may be interested in. All of our workbooks come equipped with detailed video explanations to make your learning experience a breeze! Visit us at **www.argoprep.com**

COMMON CORE MATH SERIES

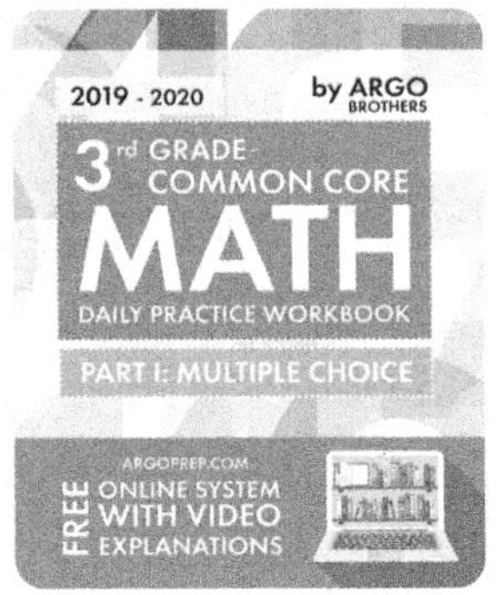

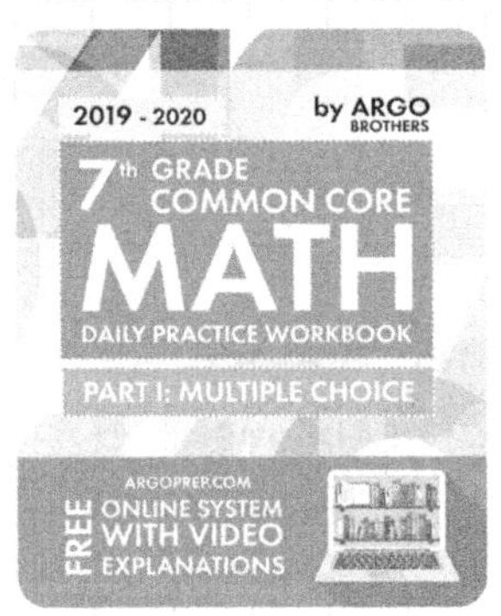

COMMON CORE ELA SERIES

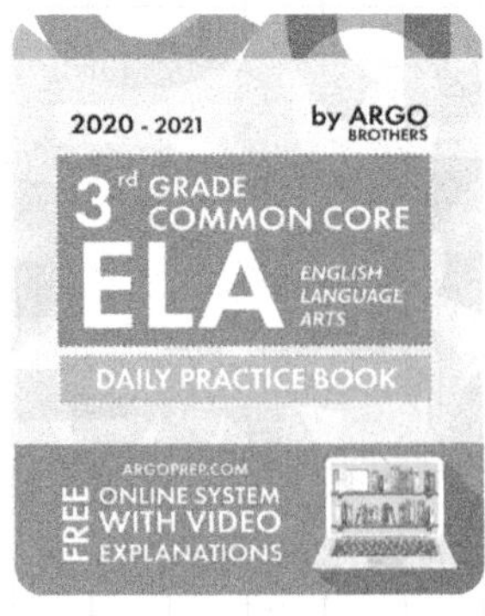

INTRODUCING MATH!

Introducing Math! by ArgoPrep is an award-winning series created by certified teachers to provide students with high-quality practice problems. Our workbooks include topic overviews with instruction, practice questions, answer explanations along with digital access to video explanations. Practice in confidence - with ArgoPrep!

YOGA MINDFULNESS FOR KIDS

HIGHER LEVEL EXAMS

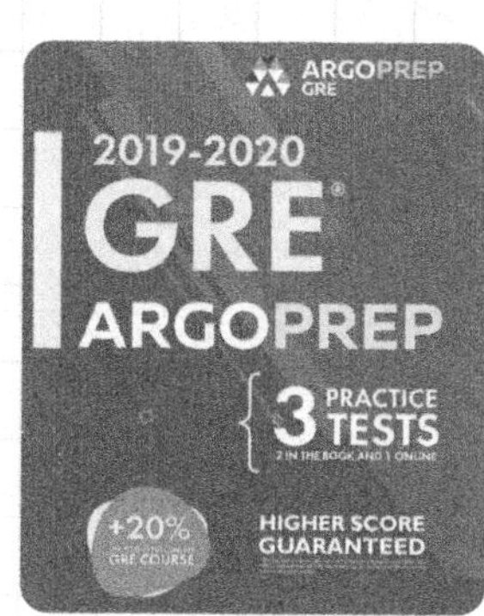

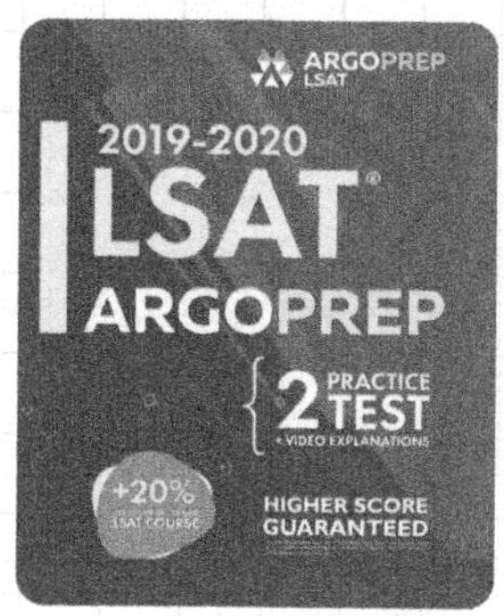

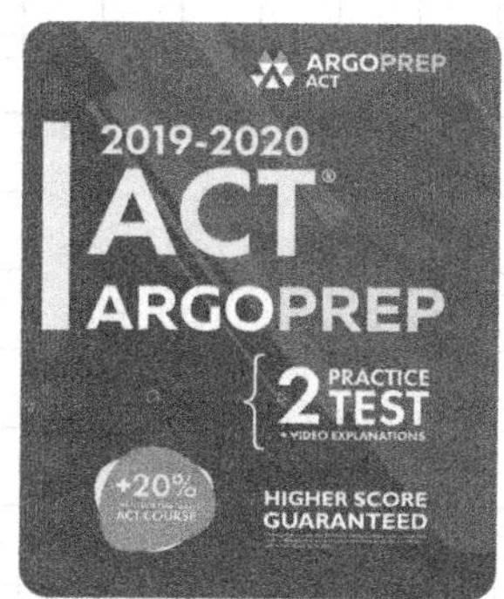

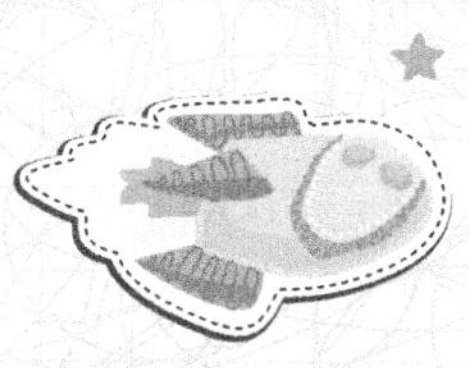

WORKBOOKS INCLUDED

Comprehensive K-8 Math & ELA Program

www.argoprep.com/k8

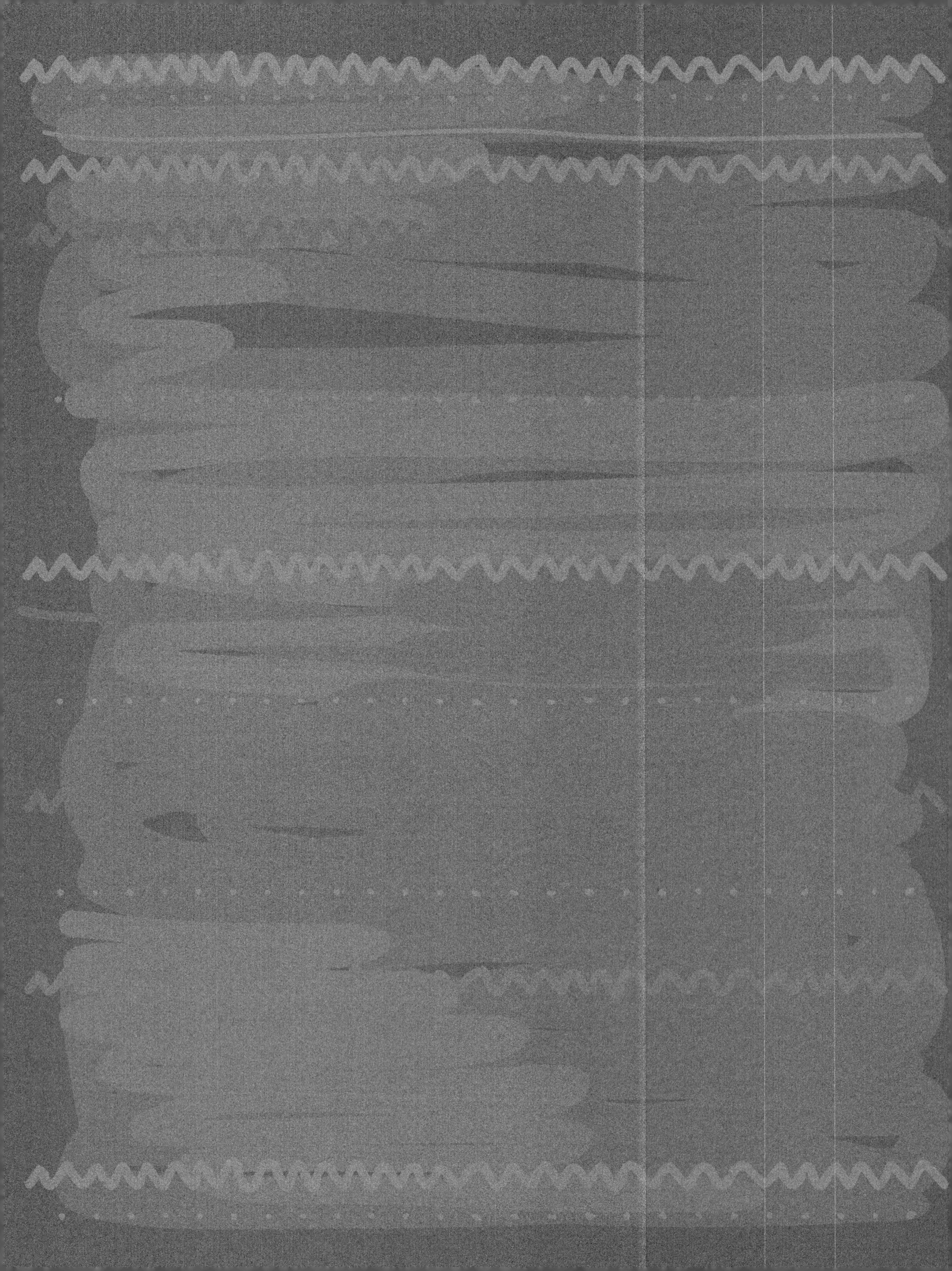

Get ready to learn about:

- parts of speech
- informational text
- simple machines
- exploring communities and more!

1. 6 + 7 = ☐

A. 12 B. 13 C. 14 D. 15

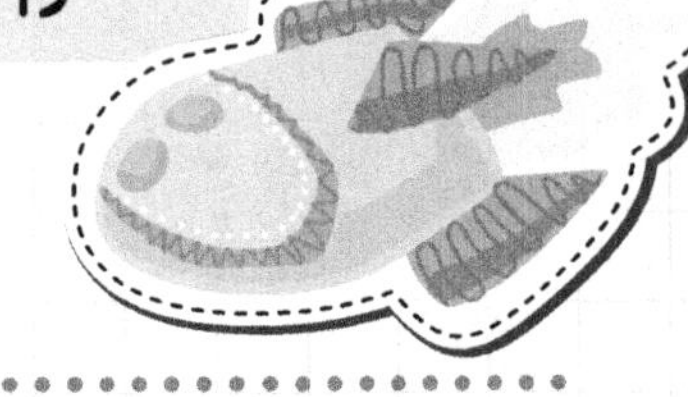

2. 4 + 9 = ☐

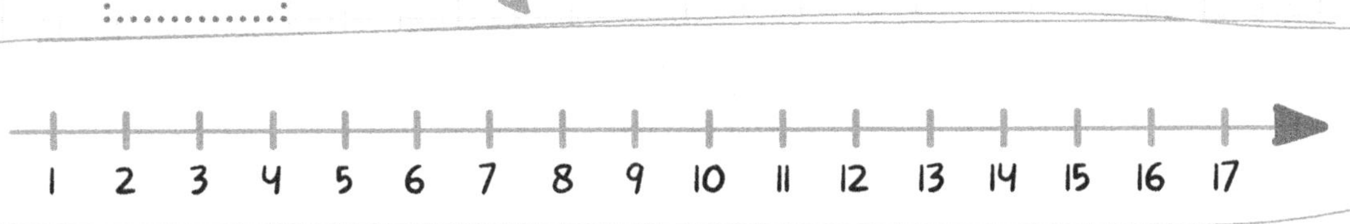

A. 13 B. 14 C. 15 D. 16

3. 2 + 5 = ☐

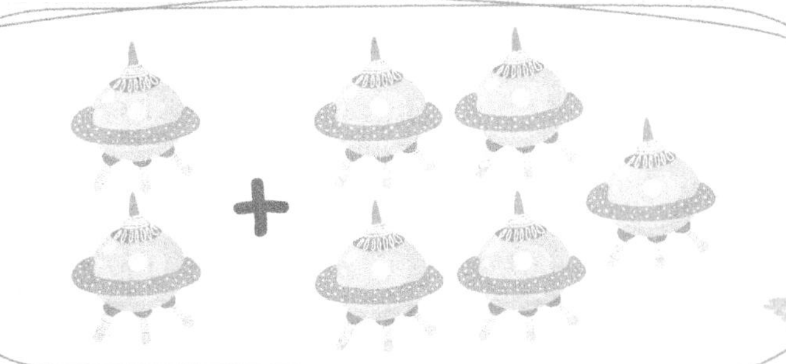

A. 5 B. 6 C. 7 D. 8

4. 1 + 3 = ☐

$$\begin{array}{r} 1 \\ +\ 3 \\ \hline \end{array}$$

A. 1 C. 3
B. 2 D. 4

5. 7 + 4 = ☐

$$\begin{array}{r} 7 \\ +\ 4 \\ \hline \end{array}$$

A. 11 C. 13
B. 12 D. 14

Week 1 Operations and Algebraic Thinking

Topic 1 Mental Math: Add or Subtract to 20

1. 11 - 7 = ☐

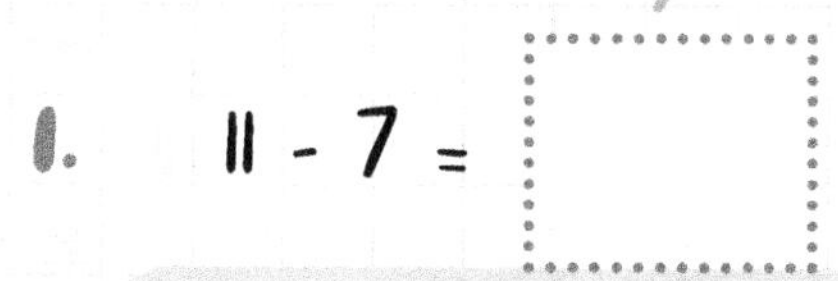

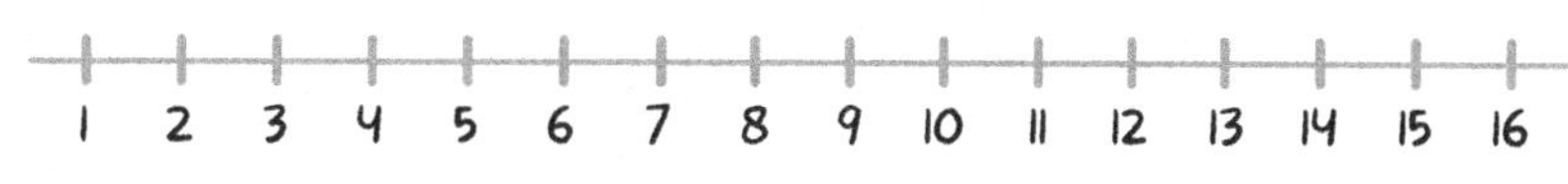

A. 4 C. 6

B. 5 D. 7

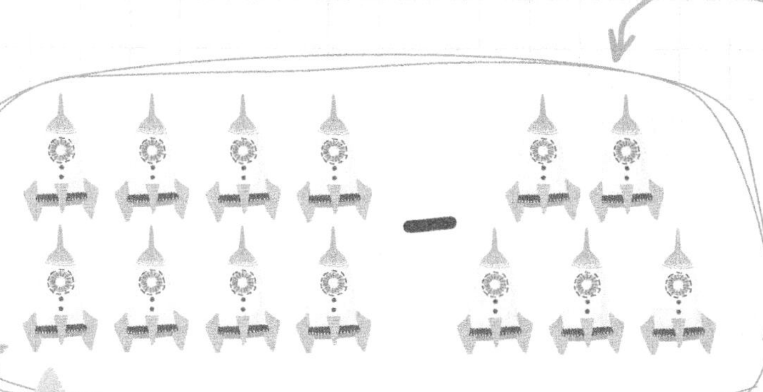

2. 8 - 5 = ☐

A. 3 B. 4 C. 5 D. 6

3. 19 - 11 = ☐

$$\begin{array}{r} 19 \\ -\ 11 \\ \hline \end{array}$$

A. 5 C. 7

B. 6 D. 8

4. 14 - 3 = ☐

$$\begin{array}{r} 14 \\ -\ 3 \\ \hline \end{array}$$

A. 8 C. 10

B. 9 D. 11

5. 6 - 2 = ☐

A. 3 C. 5

B. 4 D. 6

6. 13 - 9 = ☐

A. 4 C. 6

B. 5 D. 7

Week 1 Language Review
Topic 1 Parts of Speech

Key Vocabulary

- **Noun:** a person, place, thing or idea
  - **Examples:** teacher, mailman, Japan, state, math
- **Pronoun:** a word used in place of a noun
  - **Examples:** she, it, his, I, you
- **Verb:** an action word
  - **Examples:** talk, swing, think, write
- **Adjective:** a word that describes a noun
  - **Examples:** blue shirt, messy bedroom, seven horses
- **Adverb:** a word that describes a verb, adjective or another adverb; answers the questions how, where, when, how much and how often
  - **Examples:** quickly walked, extremely tired, often late

All of these different parts of speech work together to make a COMPLETE sentence!

Example: His mom found the blue shirt in the messy bedroom and she quickly washed it.

Nouns: mom, shirt, bedroom
Pronoun: she
Verbs: found, washed
Adjectives: his, blue, messy
Adverb: quickly

Look at the sentences below and identify the different parts of speech.

1. John ran quickly to the playground because he wanted the blue swing.

   Nouns: ....

   Pronouns: ....

   Verbs: ....

   Adjectives: ....

   Adverbs: ....

2. The black cat stalked the yarn ball mischievously before she pounced.

   Nouns: ....

   Pronouns : ....

   Verbs: ....

   Adjectives: ....

   Adverbs: ....

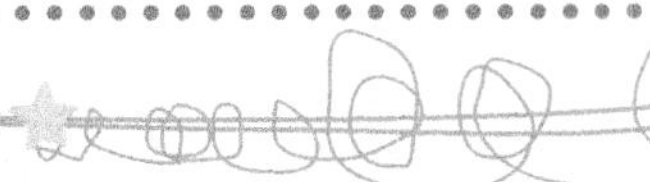

Week 1 Reading Passage

Topic 1 Informational Text

Read the passage. Then, answer the questions below.

Eating the Rainbow

Did you know there are six main food groups? They are protein, grains, dairy, fruits, vegetables and oils. It is important to eat a variety of foods from each of these food groups daily to keep your body healthy. Sometimes this is referred to as "eating the rainbow." What does that mean? It means that you should try to eat various food colors everyday. This helps your body get all the vitamins, minerals and other nutrients it needs.

Think about what you ate yesterday for breakfast, lunch, dinner and snacks. Tally how many foods you ate from each of the food groups listed above. How many different colors did you eat? If you usually eat the same fruits and vegetables everyday, challenge yourself to try a new one each week. Your body will thank you!

1. What is the main idea of this passage?
   - A. There are six main food groups.
   - B. Eating the rainbow means to eat different colored foods.
   - C. You should try new fruits and vegetables.
   - D. Eating a variety of foods from each food group helps to keep your body healthy.

2. Write one detail sentence from the passage.

....

....

3. In your own words, what does "eating the rainbow" mean?

....

....

4. Why is it important to eat a variety of foods?

....

....

5. List the six food groups.

....

....

Let's get some fitness in! Go to page 167 to try some fitness activities.

Week 1 Science

Topic 1 Investigating Simple Machines

Today you are going to learn to identify, investigate and build a simple machine. A **simple machine** is a basic mechanical device used to apply force.

There are 6 types of simple machines.

1. **Lever** - used for raising or moving a weight

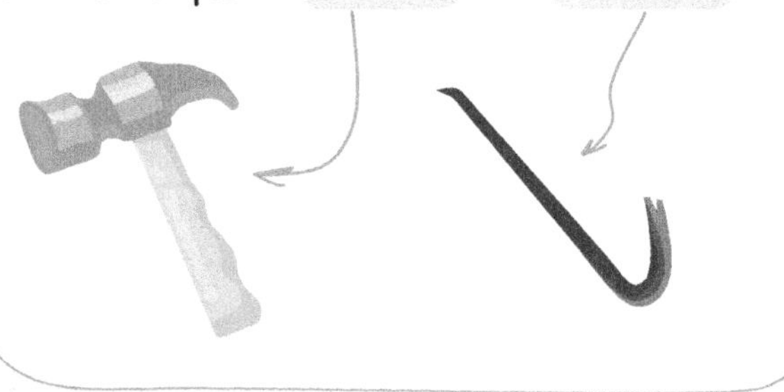

A. Example: a hammer and crowbar

2. **Inclined plane** - used for moving objects up and down more easily

A. Example: ladder

3. **Screw** - used to raise and lower things and hold things together

A. Example: screw

4. **Wheel and axle** - when the wheel turns, the axle turns too

A. Example: a bicycle tire

5. **Wedge** - used to separate things by cutting or splitting

A. Example: axe

6. **Pulley** - used for moving objects up, down or across a long distance

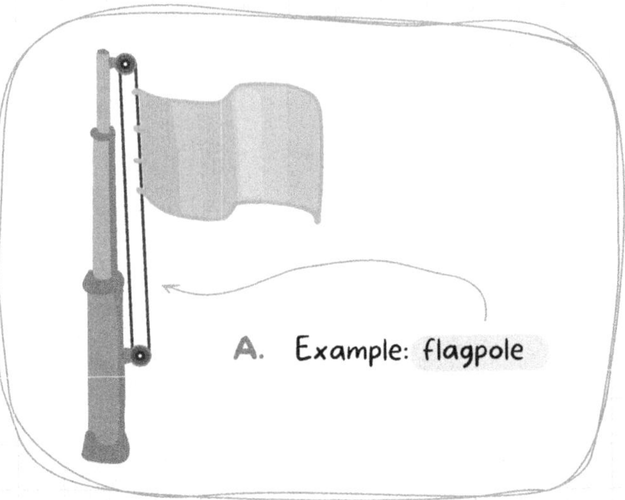

A. Example: flagpole

Week 1 Science

Topic 1 Investigating Simple Machines

Today you will build and investigate a lever by conducting the experiment below.

Materials Needed:

* Ruler
* Popsicle stick
* Large bean (such as a kidney bean)
* Pencil
* Marker (should be wider in circumference than the pencil)
* Coin

Procedure:

1. Place the pencil under the popsicle (perpendicular to each other) to create a lever.
2. Place the bean on one end of the lever.
3. Drop the coin onto the opposite side of the lever. Notice how high the bean jumps. Repeat from different heights.
4. Replace the pencil with the marker and repeat steps 2 and 3.
5. Try moving the fulcrum (pencil or marker) to different locations under the popsicle stick. Repeat steps 2 and 3.

Follow-Up Questions

How does the change from pencil to marker affect how high the bean jumps?

Does moving the fulcrum (pencil or marker) to different locations under the popsicle stick affect how high the bean jumps? How?

What caused the bean to jump the highest?

Week 1 Operations and Algebraic Thinking

Topic 2 Addition and Subtraction Word Problems

1. You have 18 cookies. You give one to each of your 6 friends. How many do you have left?

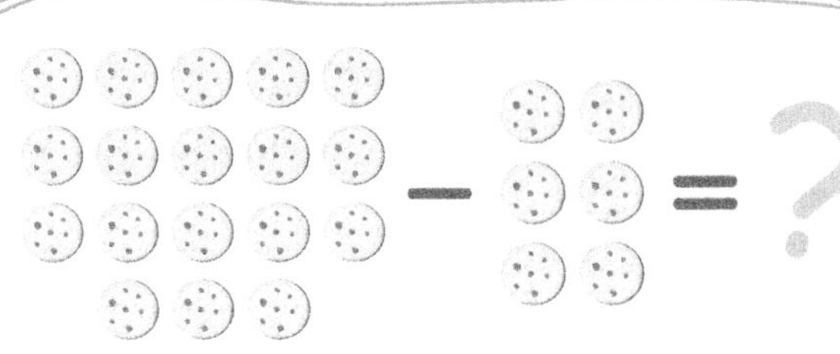

A. 12 cookies B. 11 cookies C. 10 cookies D. 9 cookies

2. You check out three books from the school library. You go to the public library and get 4 more books. How many books do you have to read in all?

A. 6 books C. 8 books
B. 7 books D. 9 books

3. You have 13 stuffed animals. You share two with your best friends. How many do you have left?

A. 10 animals C. 12 animals
B. 11 animals D. 13 animals

4. You have an apple, a banana and an orange. Your friend has two apples and 4 plums. How many pieces of fruit do you have between the two of you?

A. 7 pieces B. 8 pieces C. 9 pieces D. 10 pieces

5. Your teacher gives everyone 15 pencils to start the year. You have used 5 so far. How many pencils do you have left?

A. 8 pencils B. 9 pencils C. 10 pencils D. 11 pencils

Week 1 Operations and Algebraic Thinking

Topic 2 Mental Math: Add or Subtract to 20

1. You collect rocks. You find 4 rocks on a walk in your neighborhood. You find 6 rocks in the schoolyard. How many rocks are you going to add into your collection?

A. 10 rocks B. 11 rocks C. 12 rocks D. 13 rocks

2. Your cat loves toy mice. She has 4 mice hidden around the house. You find five more mice. How many does she have in all?

A. 8 mice B. 9 mice C. 10 mice D. 11 mice

3. Your friend brings 11 cupcakes to school. She shares them with five friends. How many cupcakes does she have left?

A. 3 cupcakes B. 4 cupcakes C. 5 cupcakes D. 6 cupcakes

4. It takes you 10 minutes to walk to the bus stop. It takes 3 minutes to get ready to leave the house. How much time should you give yourself so you are not late for your bus?

A. 8 minutes C. 11 minutes

B. 10 minutes D. 13 minutes

5. You have 15 stuffed animals. You get 10 more for your birthday. How many stuffed animals do you have now?

A. 20 animals C. 30 animals

B. 25 animals D. 35 animals

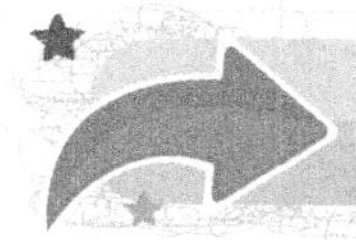

FITNESS PLANET

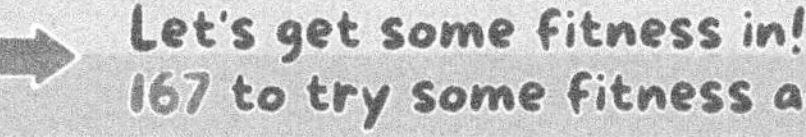

Let's get some fitness in! Go to page 167 to try some fitness activities.

Read the passage. Then, answer the questions below.

Maxwell and the Hot Day

Late one afternoon, Charlie and his mom decided to take their big, furry dog, Maxwell, for a walk. It was a very hot and sunny day so they packed water bottles and set off. As they were walking, they stopped frequently so that Charlie could examine butterflies, grasshoppers and cicadas. Maxwell liked to sniff the grass and tree trunks whenever they stopped.

After walking for close to half an hour, Charlie asked his mom for a drink of water. He was feeling hot and sweaty! They walked to a nearby bench and sat down for a break. Maxwell laid down in the shade next to them, panting. That's when they realized that Maxwell was thirsty too! Since they didn't have a bowl to put water in for him, they decided to turn around and walk back home.

When they got up to leave, Maxwell wouldn't get up from his shady spot in the grass. Charlie and his mom tried everything to get Maxwell to move, but he wouldn't budge! He was too hot and tired! What were they going to do?

Charlie thought and thought and finally he had a great idea! He set off to find the materials he needed. As he walked around, gathering items from the grass, his mom stayed with Maxwell in the shade, watching him curiously. Finally, he had all the items he needed. He was going to make a bowl for Maxwell to drink from with sticks, leaves and rocks.

Charlie knelt down on the ground and got to work. Soon, he had created a makeshift bowl out of only materials from nature. His mom carefully poured water from her bottle into the bowl. Maxwell immediately stood up from his shady spot, went over to investigate and began lapping water from the bowl. After a couple of minutes, Maxwell was happy and ready to walk back home. Charlie felt very proud of himself for solving their problem!

Read and respond to the following questions about the reading passage.

1. From which point of view is the passage being told?

   A. Charlie
   B. His mom
   C. Maxwell
   D. Narrator

2. Describe how Maxwell was feeling in the middle of the story.

3. Which word best describes Charlie?

   A. Lazy
   B. Determined
   C. Shy
   D. Tired

4. Circle the adverb in the following sentence from the passage.

   His mom carefully poured water from her bottle into the bowl.

5. List three verbs from the passage.

6. Why did Charlie create a bowl for water from materials found in nature for Maxwell?

Topic 3 Even or Odd

Indicate if this number is even or odd: 16

1. 16 EVEN ....

2. 72 ....

3. 45 ....

4. 72 ....

5. 28 ....

6. 81 ....

7. 9 ....

8. 33 ....

9. 97 ....

10. 58 ....

11. 46 ....

12. 54 ....

13. 23 ....

14. 17 ....

15. 84 ....

16. 69 ....

17. 31 ....

18. 78 ....

19. 5 ....

20. 92 ....

Week 1 Exploring Communities

Topic 1 Exploring Communities

A **community** is a group of individuals living in the same place or having a particular characteristic in common.

You are part of many different communities, some small and others much larger.

Examples of communities you may be a part of include: your family, school, neighborhood, city, state and country, among many others.

Today you are going to research your local community and answer several questions. Your local community may be your neighborhood, village or city. Use the questions below as a guide for your research.

1. What local community are you researching?

2. When was this community established?

3. How was this community established?

4. Who were the founders and early settlers?

5. What makes this community unique (think about cultural diversity, arts, architecture, industry, etc.)?

6. What is one way this community has changed over time?

7. What is one way this community has stayed the same over time?

8. Write about a specific person, event or development that brought an important change to this community.

Let's get some fitness in! Go to page 167 to try some fitness activities.

Let's train your brain for:

* building arrays
* prefixes and suffixes
* rocks and minerals
* exploring communities and more!

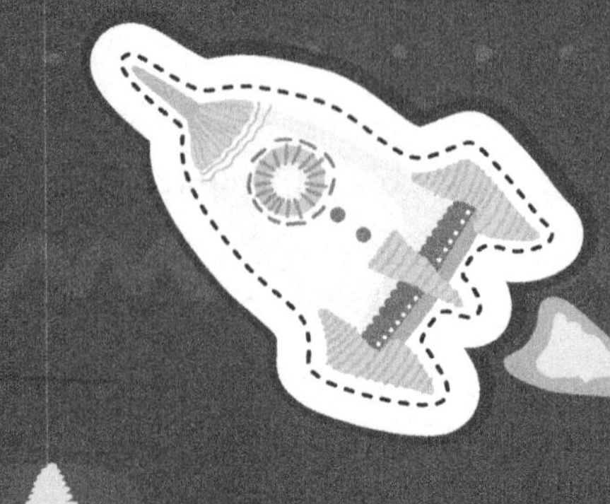

Topic 1 Building Arrays

1. How many circles are there?

....

A. 5 B. 13 C. 15 D. 3

2. How many circles are there?

....

A. 8 B. 2 C. 4 D. 10

3. How many circles are there?

....

A. 14 B. 16 C. 18 D. 20

4. How many circles are there?

....

A. 2 B. 3 C. 4 D. 5

Week 2 Operations and Algebraic Thinking

Topic 1 Building Arrays

1. How many rockets are there?

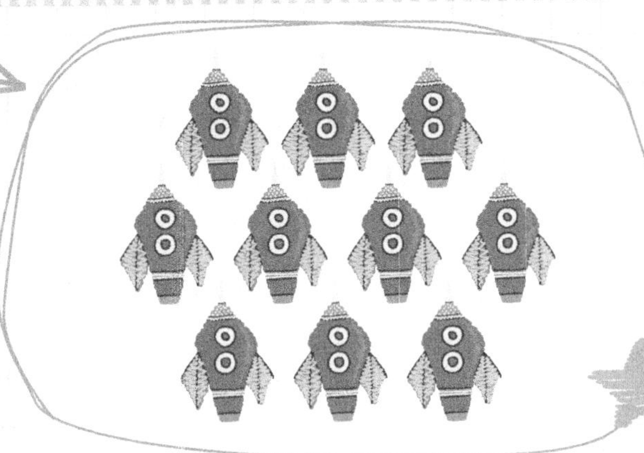

....

A. 10 B. 12 C. 14 D. 16

2. How many circles are there?

....

A. 27 B. 36 C. 45 D. 18

3. How many rockets are there? ....

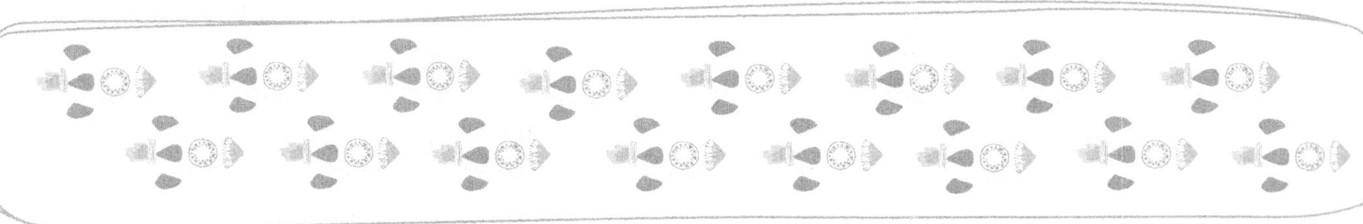

A. 16 B. 18 C. 20 D. 22

4. How many objects ware there? ....

A. 20 B. 21 C. 22 D. 23

Week 2 Language Review

Topic 1 Prefixes and Suffixes

A prefix is a group of letters attached to the beginning of a word to change its meaning.

Prefix	Meaning	Example
pre-	before	prepay
un-	not	untie
re-	again	rewrite
dis-	not, opposite of	dishonest
mis-	wrong, bad	misbehave

Example: dis- + obey = disobey, Meaning: the opposite of obey

Practice

1. re- + do = ... Meaning: ....
2. pre- + view = ... Meaning: ....
3. un- + covered = ... Meaning: ....
4. mis- + heard = ... Meaning: ....
5. dis- + cover = ... Meaning: ....
6. Write an example of a word with the prefix un- ....
7. Write an example of a word with the prefix re- ....
8. Circle the correct meaning of distrust.

   A. Not trust
   B. Trust again
   C. Trust before
   D. One who trusts

Week 2 Language Review

Topic 1 Prefixes and Suffixes

A suffix is a group of letters attached to the end of a word to change its meaning.

Suffix	Meaning	Example
-er	one who does something	teacher
-less	without	careless
-ful	full of	wonderful
-ly	how	quietly
-able	able to be	washable

Example: work + -er = worker, Meaning: one who works

Practice

1. color + -less = ... Meaning: ....
2. quick + -ly = ... Meaning: ....
3. do + -able = ... Meaning: ....
4. care + -ful = ... Meaning: ....
5. speak + -er = ... Meaning: ....
6. Write an example of a word with the suffix -ly ....
7. Write an example of a word with the suffix -able ....
8. Circle the correct meaning of colorful.
   A. Able to be colored
   B. One who colors
   C. Full of color
   D. Without color

Topic 1 Investigating Rocks and Minerals

Today you are going to observe the characteristics of rocks and minerals and learn to identify and classify rocks.

Minerals are naturally occurring substances formed by geological processes. There are over 4000 different types of minerals. Examples of minerals are: pyrite, gold, quartz and diamond.

Rocks are naturally occurring solids made up of minerals. **There are 3 types of rocks:** igneous, sedimentary and metamorphic. See the table below for how each type of rock is formed, as well as examples of each.

Igneous	Formed when magma cools and solidifies	Granite, obsidian, pumice
Sedimentary	Formed by sediment, deposited over time, at the bottom of lakes and oceans	Sandstone, mudstone, chalk
Metamorphic	Formed by extreme pressure and heat over time	Marble, slate

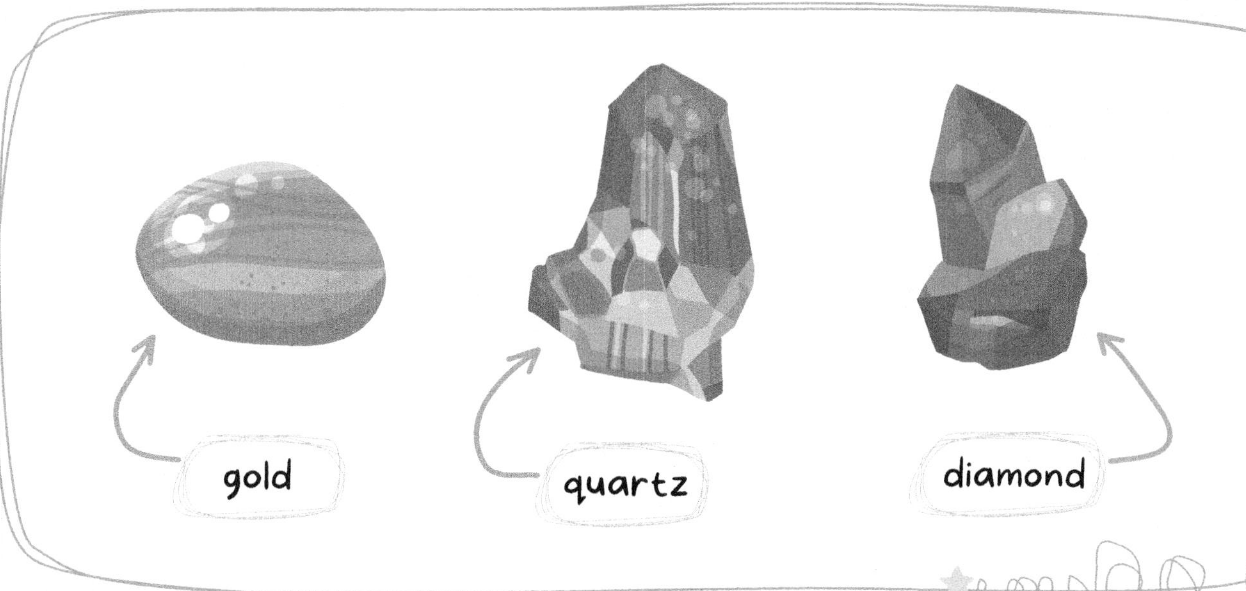

Week 2 Science

Topic 1 Investigating Rocks and Minerals

Today, you will collect rocks and perform a simple test to determine if any of your rocks are limestone.

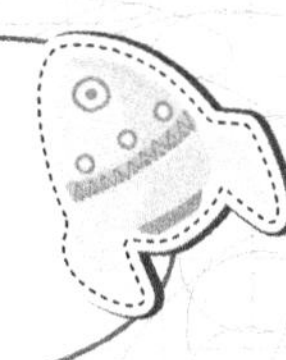

Materials Needed

* Collection of rocks
* White vinegar
* Several glass bowls (glass or plastic) or mason jars

Procedure

1. Place one rock in each bowl or jar. Cover the rock with vinegar.
2. Observe your rocks for any reactions to the vinegar.
3. Within one minute, bubbles will form around any rock that is limestone.

Follow-Up Questions

1. Did bubbles form around any of your rocks?

....

....

2. What can you determine about the types of rocks you collected?

....

....

3. Why do you think limestone reacts to an acidic liquid such as vinegar?

....

....

Week 2 Operations and Algebraic Thinking

Topic 2 Representing Arrays with Equations

1. Write an equation that represents the picture below.

2. Write an equation that represents the picture below.

3. Write an equation that represents the picture below.

4. Write an equation that represents the picture below.

5. Write an equation that represents the picture below.

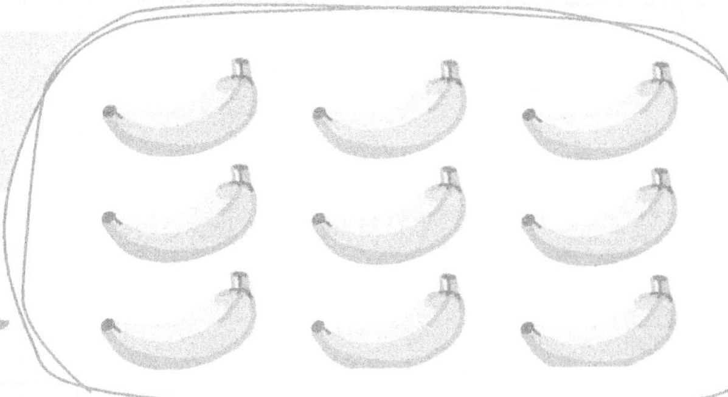

FITNESS PLANET

Let's get some fitness in! Go to page 167 to try some fitness activities.

Week 2 Operations and Algebraic Thinking

Topic 2 Representing Arrays with Equations

1. Write an equation that represents the picture.

2. Write an equation that represents the picture.

3. Write an equation that represents the picture.

4. Write an equation that represents the picture.

5. Write an equation that represents the picture.

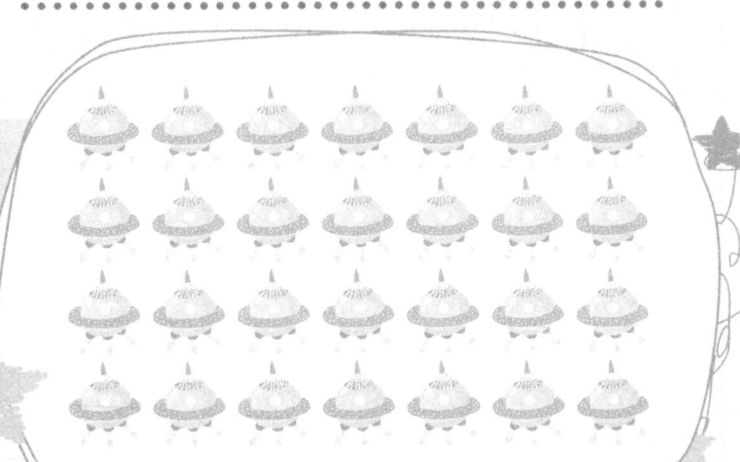

Read the fable below and then answer the questions that follow.

Narwhal and Whale lived in the same part of the deep, blue ocean. Whale was so jealous of Narwhal because she had a beautiful horn that made her unique.

"I wish I had a horn like yours," Whale told with Narwhal one day.

"I wish I could give you my horn. I don't like it!" Narwhal replied.

"You don't like it? But why?" Whale asked, feeling astonished.

"Shark and his friends always make fun of my horn, and I feel embarrassed," said Narwhal.

At that very moment, Shark and several of his friends swam over and began laughing and making fun of Narwhal. Narwhal turned bright pink and looked very sad. Whale felt so sorry for Narwhal.

"Shark, why are you making fun of Narwhal?" asked Whale.

"Because she looks so silly with that horn," replied Shark, laughing meanly.

"She does NOT look silly. She looks beautiful and unique. I think you're just jealous of her and that's why you make fun of her," Whale said.

"Jealous? I'm n-not jealous," Shark replied, but he sounded unsure as he and his friends swam away.

Once they were gone, Whale went over to comfort Narwhal.

"I'm so sorry, Narwhal. Now I understand why you don't like your horn. I didn't know you got made fun of for it. I want you to know, though, that I love it and think it makes you unique!"

"Thank you," responded Narwhal, as they swam off to play.

This story is a fable. A **fable** is a short story that usually has animals as characters and teaches the reader a **moral**, or lesson.

The moral of this story is to be thankful for what you have.

1. What is another possible moral for this story?

....

2. Why do you think Shark sounded unsure at the end of the story?

....

Now, you give it a try. Think of a moral or lesson like the one above and write a fable about it.

Week 2 Reading Passage
Topic 3 Informational Text

Read the passage below and then answer the questions that follow.

Table of Contents

Where Koalas Live

Koalas are **marsupials,** a type of mammal, that are native to Australia. They are found mainly in the coastal regions of the continent. Koalas usually live in **eucalyptus** woodland areas since the leaves of that tree make up a large portion of their diet.

This passage contains several examples of **text features.** Text features are components of a text that provide the reader information. Examples of text features include: table of contents, bold text, captions, headings, glossary and labeled diagrams.

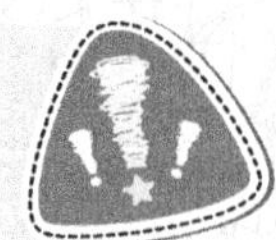

1. List 3 different text features that are used in this passage.
2. In what chapter would you find information about what koalas look like?
3. In what chapter would you find information about eucalyptus trees?
4. What type of information might be found in Chapter 1?
5. Why are the words "marsupials" and "eucalyptus" in bold text in Chapter 2?
6. What is an example of a text feature that could be added to this passage to help the reader?

Now, conduct an online search about koalas and write 3 things you learned about them.

Draw an array that represents:

2. 2 + 2

1. 4 + 4 + 4

3. 3 + 3 + 3 + 3

4. 1 + 1 + 1

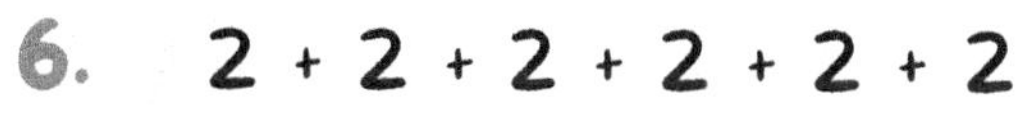

6. 2 + 2 + 2 + 2 + 2 + 2

5. 5 + 5

FITNESS PLANET

Let's get some fitness in! Go to page 167 to try some fitness activities.

Immigration is the act of going to live permanently in a foreign country.

Ellis Island is an immigration station located between New York and New Jersey that was in operation from 1892-1954. Millions of immigrants from all over the world passed through Ellis Island.

Today you are going to research Ellis Island more in depth and answer the following questions.

1. During its peak years of operation, how many people passed through Ellis Island each day?
2. What makes Ellis Island an important part of our country's history?
3. Why is Ellis Island still an important landmark today?
4. Write 3 facts you learned about Ellis Island.
5. Thinking back to the community you researched last week, how did immigration play a role in that community? Where did most of the community's early settlers emigrate from?
6. How do you think immigration enriches a community?

It's time to explore:

- hundreds, tens and ones
- skip counting
- plural nouns
- exploring fossils and more!

Week 3 Operations and Algebraic Thinking

Topic 1 Place value: Hundreds, Tens and Ones

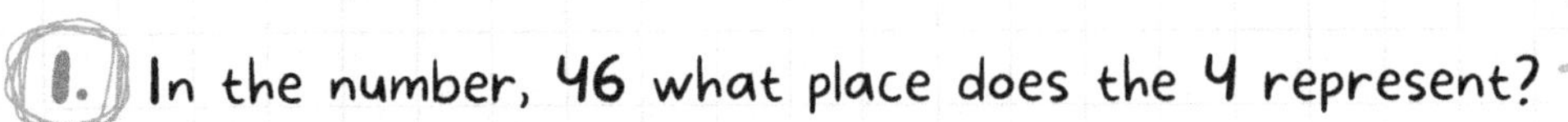

1. In the number, 46 what place does the 4 represent?

A. Ones B. Tens C. Hundreds D. Zeros

2. In the number, 298 what place does the 2 represent?

A. Zeros B. Hundreds C. Tens D. Ones

3. In the number, 73 what place does the 3 represent?

A. Ones B. Tens C. Hundreds D. Zeros

4. In the number, 104 what place does the 0 represent?

A. Zeros B. Hundreds C. Tens D. Ones

5. In the number, 328 what place does the 2 represent?

A. Ones B. Tens C. Hundreds D. Zeros

6. In the number, 81 what place does the 8 represent?

A. Zeros B. Hundreds C. Tens D. Ones

7. In the number, 960 what place does the 9 represent?

A. Ones B. Tens C. Hundreds D. Zeros

Topic 1 Place value: Hundreds, Tens and Ones

1. In the number, 37 what place does the 3 represent?

A. Ones B. Tens C. Hundreds D. Zeros

2. In the number, 144 what place does the 1 represent?

A. Zeros B. Hundreds C. Tens D. Ones

3. In the number, 86 what place does the 6 represent?

86

A. Ones B. Tens C. Hundreds D. Zeros

4. In the number, 909 what place does the 0 represent?

A. Zeros B. Hundreds C. Tens D. Ones

5. In the number, 521 what place does the 5 represent?

521

A. Ones B. Tens C. Hundreds D. Zeros

6. In the number, 835 what place does the 8 represent?

835

A. Zeros B. Hundreds C. Tens D. Ones

FITNESS PLANET

Let's get some fitness in! Go to page 167 to try some fitness activities.

Week 3 Language Review

Topic 1 Plural Nouns

Plural means more than one. 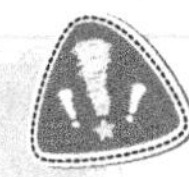

A **noun** is a person, place or thing. 

There are several ways to change a singular noun into a regular plural noun.

Add -s	Add -es	Drop the -y, add -ies	Drop the -f, add -ves
dog → dogs fruit → fruits shoe → shoes	bush → bushes lunch → lunches miss → misses	story → stories family → families lady → ladies	wolf → wolves leaf → leaves shelf → shelves

Irregular nouns do not follow the rules when changing into plural nouns.

tooth → teeth	deer → deer	child → children	man → men

Practice changing the following singular nouns into plural nouns.

1. color ....
2. baby ....
3. flash ....
4. knife ....
5. vegetable ....
6. fish ....
7. woman ....
8. Write an example of an irregular plural noun. ....

....

....

Week 3 Reading Passage
Topic 2 Literature

Authors often use figurative language to make their writing more dramatic or interesting to readers. It is important to be able to tell the difference between literal and non-literal text to better understand it.

Literal text: the text means exactly what it says.

Example: While playing with cars, the little boy kept driving them up the wall!

Explanation: The little boy is literally driving cars up the wall.

Non-literal text (also called **figurative language**): the text does not mean exactly what it says; uses words to achieve a more dramatic effect.

Example: My mom used to tell us, "You're driving me up the wall!"

Explanation: Mom is not literally driving up the wall. This phrase is used as a dramatic way of saying the kids were making her feel annoyed or angry.

Read the passage below.

On Friday, Jason had a social studies test during third period. The night before, he hadn't bothered studying because he thought the test would be a piece of cake. When he arrived to class on Friday, Jason began to feel nervous. Once his teacher passed out the test papers, Jason got to work.

When the bell rang, Jason walked out of the classroom with a sinking feeling. He knew he hadn't done well and that he should've studied instead of playing video games the night before.

As Jason walked home from school later that day, he imagined bringing home his test the following week and his mom chewing him out for the poor grade. It was going to be a long weekend of waiting and worrying!

Week 3 Reading Passage

Topic 2 Literature

Now you will practice by answering the questions below.

1. List two examples of figurative language from the passage.

2. What type of text is "Jason had a social studies test during third period?"
   A. Literal text
   B. Non-literal text

3. Why did Jason have a sinking feeling as he walked out of the classroom?

4. Why did the author use figurative language in this passage?

5. Write your own example of figurative language.

6. Use your example of figurative language in a sentence.

7. Write your own example of literal language and use it in a sentence.

8. What is the hidden meaning of "break a leg?"

9. What is the hidden meaning of "zip your lips?"

10. Write a well-constructed paragraph, using at least two examples of figurative language.

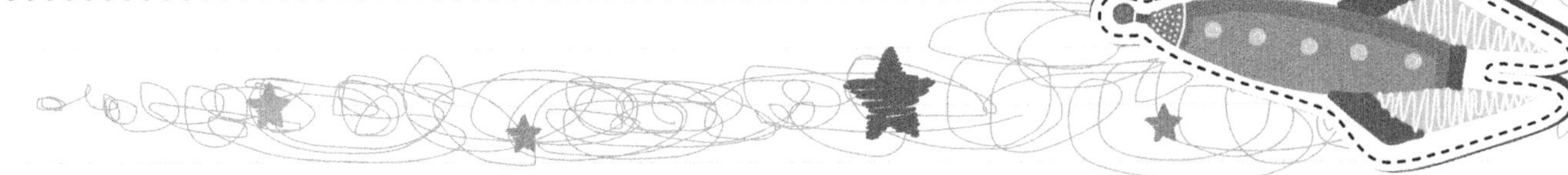

Topic 1 Exploring Fossils

A **fossil** is the remains or impression of a prehistoric organism preserved in petrified form in rock. Fossils are most commonly found in limestone, sandstone and shale.

Read the following passage about how fossils are formed.

Fossils are formed when an animal dies and its body sinks to the bottom of a body of water. The remains begin to rot and become covered in mud. Over time layers of sediments cover the bones of the animal. These sediments eventually harden into rock. The animal's bones slowly decay and minerals fill in the space where the bones were. The fossil has formed but is covered beneath many layers of rock. Over time these layers of rock get worn away by wind and rain (a process called erosion), bringing them to the surface and making them visible. The fossils are then often found.

Week 3 Science

Topic 1 Exploring Fossils

Now you will conduct an experiment to better understand how layers of sediments build up over the bones of an animal to help create fossils.

Materials Needed

- Slices of three different types of bread (it's helpful if they are different colors)
- Gummy worms or fish
- Paper towels
- Several heavy books

Procedure

1. On a paper towel, stack the three slices of bread on top of one another.
2. Insert the gummy animals into the center of the bread.
3. Cover with another paper towel and then stack the heavy books on top.
4. Wait 24 hours! No peeking!

Follow-Up Questions

1. Make a prediction about what you think will happen over the next 24 hours.

....

....

2. After 24 hours, uncover your bread slices. Record your observations.

....

....

3. What do you notice about the 3 slices of bread?

....

....

4. How does this experiment depict the process of a fossil being created?

....

....

FITNESS PLANET

Let's get some fitness in! Go to page 167 to try some fitness activities.

Topic 2 Counting and Skip Counting

10 → 11 → 12 → ?

1. What number comes after 12?

A. 10 B. 11 C. 12 D. 13

2. What number comes after 15 when counting by 5s?

A. 20 B. 25 C. 30 D. 35

3. What number comes after 17?

A. 16 B. 18 C. 20 D. 22

4. What number comes after 230 when counting by 10s?

A. 220 B. 240 C. 250 D. 280

5. What number comes after 60 when counting by 5s?

A. 65 B. 70 C. 75 D. 80

6. What number comes after 24?

A. 26 B. 25 C. 23 D. 22

7. What number comes after 83?

A. 81 B. 82 C. 84 D. 85

8. What number comes after 45 when counting by 5s?

A. 50 B. 55 C. 60 D. 65

Week 3 Operations and Algebraic Thinking

Topic 3 Reading and Writing Numbers to 1000

1. Represent 42 in words. ....

....

2. Represent 821 in expanded form. ....

....

3. Represent thirty-six using base-ten numerals. ....

....

4. Represent 231 in words. ....

....

5. Represent fifty-nine in expanded form. ....

....

6. Represent 400 + 80 + 6 in using base-ten numerals. ....

....

7. Represent 900 + 70 + 5 in using base-ten numerals. ....

....

8. Represent 307 in words. ....

....

9. Represent 674 in expanded form. ....

....

Week 3 Reading Passage

Topic 3 Informational Text

Read the passage.

The leaves are slowly changing. Green to yellow to orange to red. The cool, gentle breeze sways through the trees, and the leaves begin to shower to the ground. The people below pull their jackets tighter as they walk quickly through the park, leaves crunching underfoot.

Dusk threatens as the sun sinks lower in the western sky. The days are getting shorter. The street lights flicker on and children scramble to get home. Lights twinkle from houses and the smell of apple pie wafts through open windows. This is autumn.

Answer the following questions about the passage.

1. Circle the words in the text that tell the reader where the passage takes place.

2. Underline the words in the text that tell the reader when the passage takes place.

3. Why do you think the people are pulling their jackets tighter as they walk through the park?

4. What does the phrase "the leaves begin to shower to the ground" mean?

5. Choose your favorite season and write a similar passage to describe it.

FITNESS PLANET

Let's get some fitness in! Go to page 167 to try some fitness activities.

Topic 3 Reading and Writing Numbers to 1000

1. Represent six hundred forty-two in using base-ten numerals. ....

....

2. Represent 94 in words. ....

....

3. Represent 407 in expanded form. ....

....

4. Represent 825 in expanded form. ....

....

5. Represent 51 in words. ....

....

6. Represent twenty-one in using base-ten numerals. ....

....

7. Represent 739 in words. ....

....

8. Represent 584 in expanded form. ....

....

9. Represent 80+3 in using base-ten numerals. ....

....

Topic 3 Levels of Government

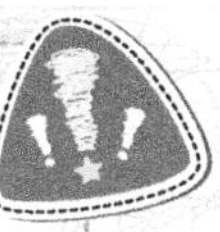

There are three levels of government - **local, state and federal**. Each level is responsible for different duties, although there is some overlap between the levels. Find out more in the table below.

Local Government	State Government	Federal Government
Road maintenance Trash collection Electricity, gas and water Libraries Schools Parks and recreation facilities First responders Polling places	Department of Motor Vehicles State licensing exams Hospitals Schools Constitutional amendment ratification State parks and forests Correctional facilities Road maintenance	Polling places Correctional facilities National parks Road maintenance Federal Aviation Administration National security Endangered species Money Health

Answer the questions below.

1. Write a short response outlining at least 3 reasons why you think it's important to have different levels of government.

2. List one other duty each level of government is responsible for.

3. Why do you think some duties fall under multiple levels of government (example: road maintenance)?

Federal Government

State Government

Local Government

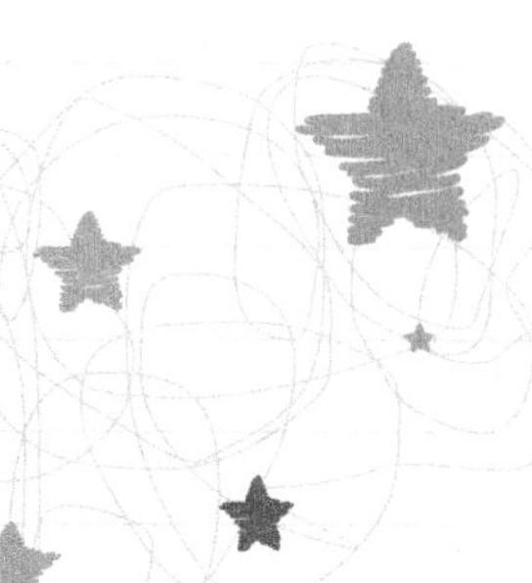

Let's see what you know about:

- comparing numbers
- opinion writing
- weather patterns
- good/services and more!

Topic 1 Comparing Numbers

1. Which number is largest?

....

A. 86 B. 68 C. 608 D. 860

2. Which number is largest?

....

A. 320 B. 230 C. 203 D. 32

3. Which number is largest?

....

A. 18 B. 1800 C. 810 D. 180

4. Which number is largest?

....

A. 41 B. 416 C. 4016 D. 610

5. Which number is largest?

....

A. 570 B. 507 C. 705 D. 75

6. Which number is smallest?

....

A. 8 B. 80 C. 800 D. 18

7. Which number is smallest?

....

A. 221 B. 210 C. 2100 D. 201

8. Which number is smallest?

....

A. 78 B. 72 C. 79 D. 80

FITNESS PLANET

Let's get some fitness in! Go to page 167 to try some fitness activities.

What symbol completes the number sentence?

1. 11 ○ 58
2. 74 ○ 31
3. 49 ○ 75
4. 97 ○ 22
5. 62 ○ 90

6. 36 ○ 21
7. 70 ○ 41
8. 63 ○ 53
9. 28 ○ 96
10. 48 ○ 85

Week 4 Writing Review

Topic 1 Opinion Writing

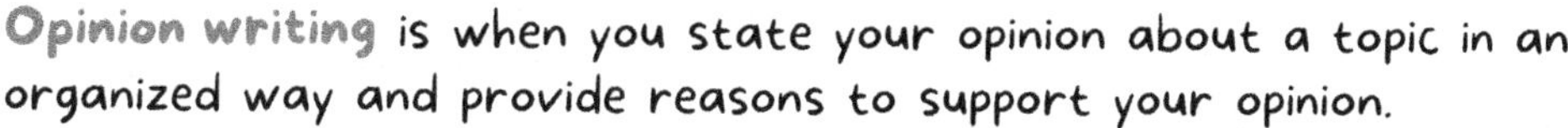

Opinion writing is when you state your opinion about a topic in an organized way and provide reasons to support your opinion.

Today you will practice writing a short opinion piece. This organized paragraph on a specific topic will include an opinion statement, reasons to support your opinion and a concluding statement. You will begin by mapping out your writing first and then put it together into a well-organized paragraph.

Think about the last book you read. This book will be your topic for this writing piece. A topic is what you are writing about. You will always start any type of writing with a topic and often use it to introduce or begin your piece of writing.

Topic: ....

What was your opinion of the book? Did you like it or not? Your opinion should be explicitly stated at the beginning of your writing piece.

Opinion: ....

Think of at least 3 reasons why you liked or disliked the book. You will use these as supporting details in your writing. You should summarize each reason in a sentence or two.

Reason #1: ....

....

....

Reason #2: ....

....

....

Reason #3: ....

....

....

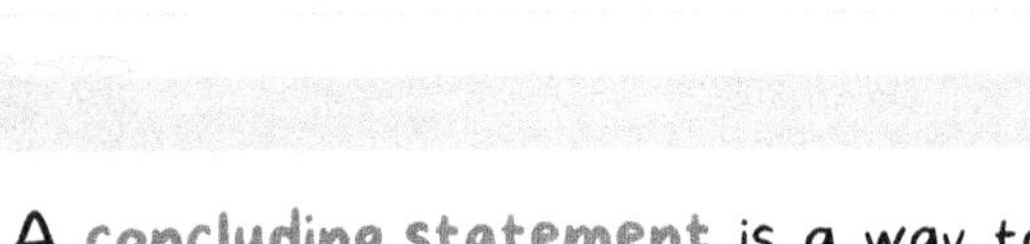

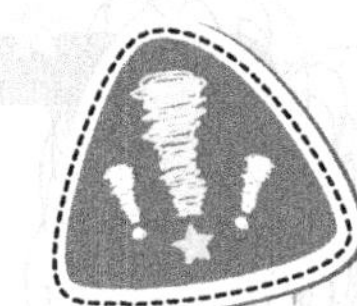

A **concluding statement** is a way to wrap up your writing. It should sum up your opinion and give the reader an idea of whether or not you recommend something or what you think should be done next. Think about whether or not you would recommend this book to others. Why or why not?

Concluding statement: ....

Now you will use the graphic organizer you created above to write a well-organized opinion piece about the last book you read on the lines below. Be sure your writing includes an opinion statement, at least 3 reasons to support your opinion and a concluding statement. Remember to write in complete sentences and use capital letters and correct punctuation.

Topic 1 Investigating Weather Patterns

The United States is divided into **seven regions**, or areas with specific characteristics. These regions are: the Northwest, West, Southwest, Midwest, Southeast, Mid-Atlantic and Northeast. These regions can be defined by certain characteristics such as geography and climate.

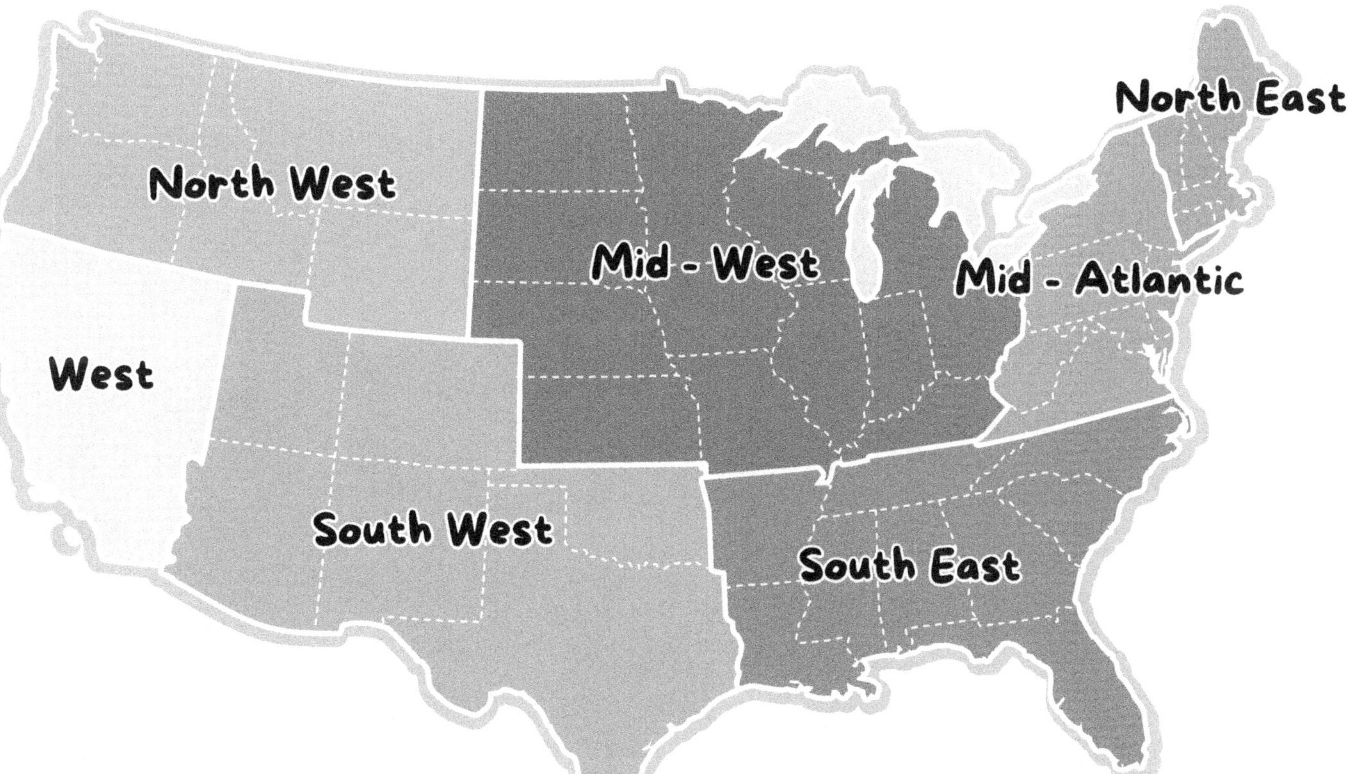

Let's get some fitness in! Go to page 167 to try some fitness activities.

Week 4 Science

Topic 1 Investigating Weather Patterns

Today, you will be researching the weather patterns, or climate, of each of the seven regions. You will use the information you find to fill out the chart below. Use the internet to research weather patterns for these regions.

	NW	W	SW	MW	SE	MA	NE
Average High Temp in Summer							
Average Low Temp in Summer							
Average High Temp in Winter							
Average Low Temp in Winter							
Average Annual Rainfall							
Average Annual Snowfall							

Topic 2 Adding Two Digit Numbers

1. 16 + 36 = ☐

$$\begin{array}{r} 16 \\ +\,36 \\ \hline \end{array}$$

A. 25 B. 52 C. 63 D. 62

2. 41 + 65 = ☐

$$\begin{array}{r} 41 \\ +\,65 \\ \hline \end{array}$$

A. 160 B. 61 C. 106 D. 161

3. 74 + 21 = ☐

$$\begin{array}{r} 74 \\ +\,21 \\ \hline \end{array}$$

A. 95 B. 105 C. 110 D. 115

4. 28 + 80 = ☐

$$\begin{array}{r} 28 \\ +\,80 \\ \hline \end{array}$$

A. 102 B. 104 C. 106 D. 108

5. 93 + 53 = ☐

$$\begin{array}{r} 93 \\ +\,53 \\ \hline \end{array}$$

A. 146 B. 148 C. 150 D. 152

6. 57 + 92 = ☐

$$\begin{array}{r} 57 \\ +\,92 \\ \hline \end{array}$$

A. 147 B. 149 C. 151 D. 153

7. 60 + 17 = ☐

$$\begin{array}{r} 60 \\ +\,17 \\ \hline \end{array}$$

A. 57 B. 67 C. 77 D. 87

Topic 3 Adding up to 1000

1. 141 + 528 = ☐

$$\begin{array}{r} 141 \\ +\ 528 \\ \hline \end{array}$$

....

....

A. 649 B. 659 C. 669 D. 679

2. 568 + 66 = ☐

$$\begin{array}{r} 568 \\ +\ 66 \\ \hline \end{array}$$

....

....

A. 633 B. 634 C. 635 D. 636

3. 822 + 265 = ☐

$$\begin{array}{r} 822 \\ +\ 265 \\ \hline \end{array}$$

....

....

A. 1087 B. 187 C. 1807 D. 1870

4. 21 + 404 = ☐

$$\begin{array}{r} 21 \\ +\ 404 \\ \hline \end{array}$$

....

....

A. 425 B. 427 C. 429 D. 431

5. 683 + 94 = ☐

$$\begin{array}{r} 683 \\ +\ 94 \\ \hline \end{array}$$

....

....

A. 444 B. 555 C. 666 D. 777

6. 924 + 19 = ☐

$$\begin{array}{r} 924 \\ +\ 19 \\ \hline \end{array}$$

....

....

A. 934 B. 955 C. 943 D. 944

7. 39 + 773 = ☐

$$\begin{array}{r} 39 \\ +\ 773 \\ \hline \end{array}$$

....

....

A. 808 B. 810 C. 811 D. 812

8. 435 + 137 = ☐

$$\begin{array}{r} 435 \\ +\ 137 \\ \hline \end{array}$$

....

....

A. 572 B. 574 C. 576 D. 578

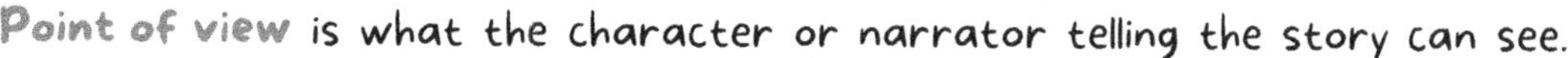

Point of view is what the character or narrator telling the story can see.

There are several types of point of view. See the chart below for more information about each of the different types.

First person	Story is being told from the point of view of a character in the story.
Second person	The narrator is speaking to YOU, the reader.
Third person limited	A narrator is telling the story and can only describe the events happening before his/her eyes.
Third person omniscient	A narrator is telling the story and can describe what each character is doing and/or thinking.
Third person limited-omniscient	A narrator is telling the story and can describe what ONE character is doing and/or thinking.

Read the passage below.

It was refreshingly cool on an evening in early autumn. I took my book out to the porch to read while the kids played with the neighbors next door. As I sat reading, with cars passing by on the street every few minutes, I could hear the kids hosting a talent competition with one another. They were singing and dancing, laughing and occasionally arguing with one another.

After about an hour, with the sun beginning to set, I yelled to the kids to come inside.

"But mom," argued Oliver, "we're having so much fun! Just a few more minutes please?"

I agreed to give them ten more minutes.

"Thanks mom," Oliver exclaimed, hugging me and running back to his friends next door.

FITNESS PLANET

Let's get some fitness in! Go to page 167 to try some fitness activities.

Now answer the following questions.

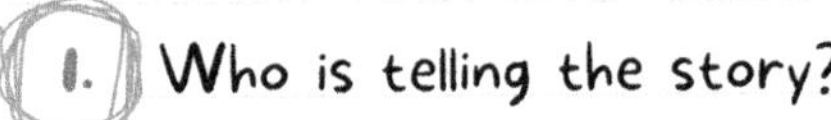

1. Who is telling the story?

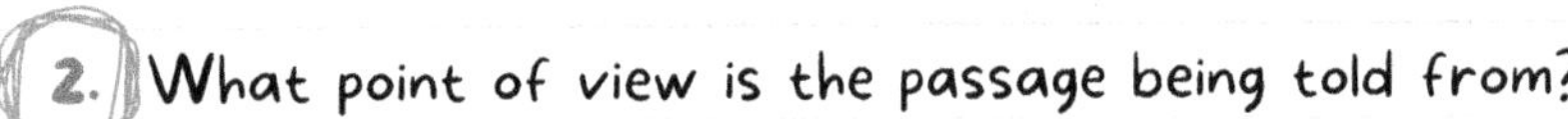

2. What point of view is the passage being told from?

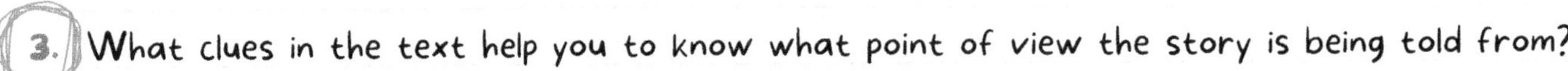

3. What clues in the text help you to know what point of view the story is being told from?

4. If a narrator were telling the story and knew what both mom and Oliver were thinking, what point of view would this be?

Now it's your turn to apply what you've learned about point of view. Write a paragraph using any of the third person point of view types.

Identify the point of view you will be writing from:

Topic 3 Adding up to 1000

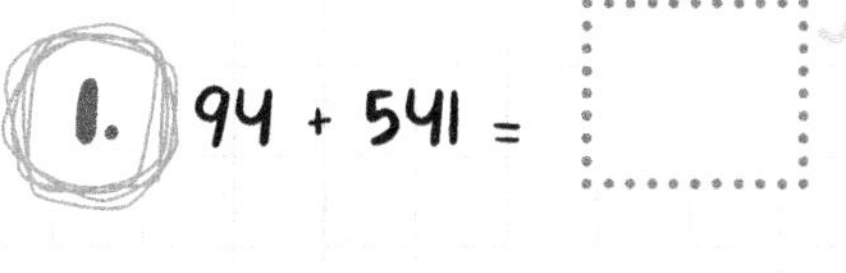
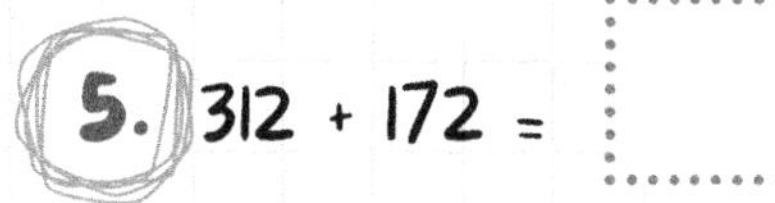

1. 94 + 541 = ☐

$$\begin{array}{r} 94 \\ +\ 541 \\ \hline \end{array}$$

A. 630 B. 633 C. 635 D. 637

5. 312 + 172 = ☐

$$\begin{array}{r} 312 \\ +\ 172 \\ \hline \end{array}$$

A. 481 B. 482 C. 483 D. 484

2. 609 + 335 = ☐

$$\begin{array}{r} 609 \\ +\ 335 \\ \hline \end{array}$$

A. 942 B. 944 C. 946 D. 948

6. 576 + 103 = ☐

$$\begin{array}{r} 576 \\ +\ 103 \\ \hline \end{array}$$

A. 677 B. 679 C. 681 D. 681

3. 423 + 90 = ☐

$$\begin{array}{r} 423 \\ +\ 90 \\ \hline \end{array}$$

A. 513 B. 515 C. 517 D. 519

7. 655 + 264 = ☐

$$\begin{array}{r} 655 \\ +\ 264 \\ \hline \end{array}$$

A. 717 B. 818 C. 919 D. 1001

4. 297 + 82 = ☐

$$\begin{array}{r} 297 \\ +\ 82 \\ \hline \end{array}$$

A. 379 B. 381 C. 383 D. 385

Topic 1 Government

Governments are responsible for providing a variety of goods and services to its citizens. Local, state and federal governments have a duty to provide different goods and services.

Goods are items that are tangible, such as books, pens, shoes and hats. Services are intangible and include things such as doctors, dentists and electricians.

Today you are going to practice identifying goods vs. services and research specific goods and services provided by the different levels of government.

1. Which of the options below are examples of goods? Circle all that apply.
   - A. Clothing
   - B. Landscaper
   - C. Furniture
   - D. Car
   - E. Plumber

2. Which of the options below are examples of services? Circle all that apply.
   - A. Pencil
   - B. Dry Cleaning
   - C. Public Transportation
   - D. Toy
   - E. Dog

3. Give another example of a good.

4. Give another example of a service.

5. What are 3 examples of goods that governments are responsible for providing to its citizens?

6. What are 3 examples of services that governments are responsible for providing to it citizens?

Grade 2-3

WEEK 5

Get ready for a new adventure with:

* mental math
* temporal words
* weather hazards
* informational texts and more!

Topic 1 Mental Math: Adding and Subtracting by 10

1. 20 + 30 = ☐

A. 40 B. 50 C. 60 D. 70

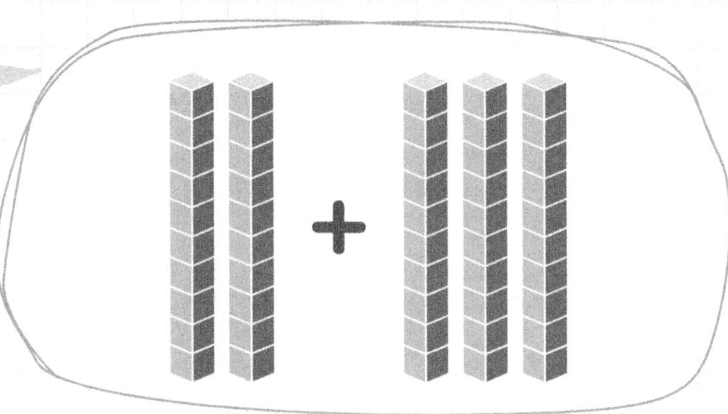

2. 50 + 40 = ☐

A. 90 B. 80 C. 70 D. 60

3. 80 + 20 = ☐

A. 70 B. 80 C. 90 D. 100

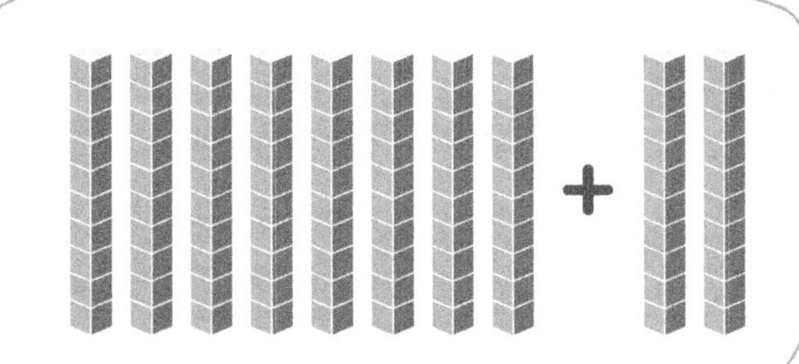

4. 90 + 40 = ☐

A. 130 B. 140 C. 150 D. 160

5. 30 + 10 = ☐

A. 30 B. 40 C. 50 D. 60

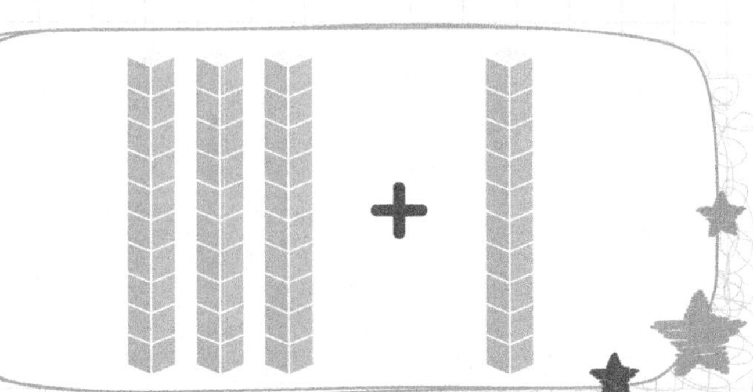

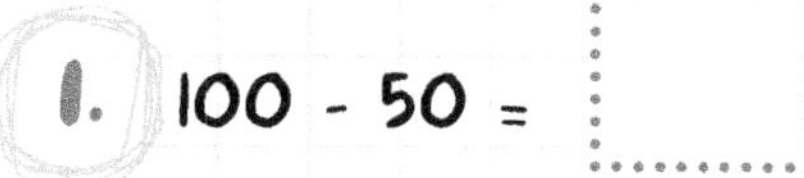

1. 100 - 50 = ☐

A. 40 B. 50 C. 60 D. 70

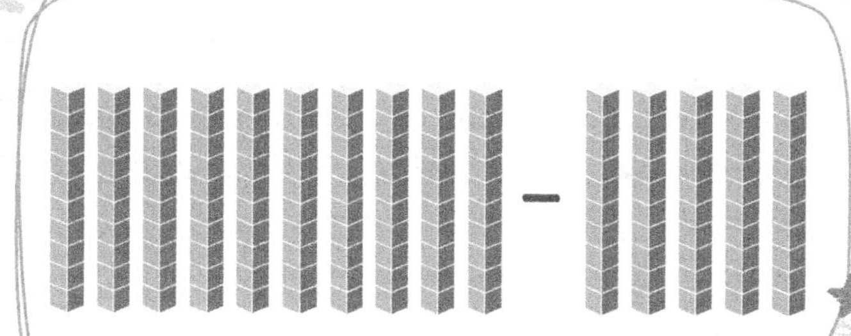

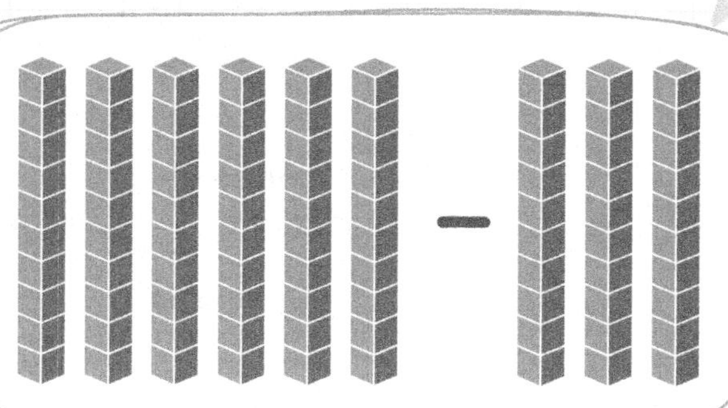

2. 60 - 30 = ☐

A. 30 B. 40 C. 50 D. 60

3. 80 - 80 = ☐

A. 0 B. 10 C. 20 D. 30

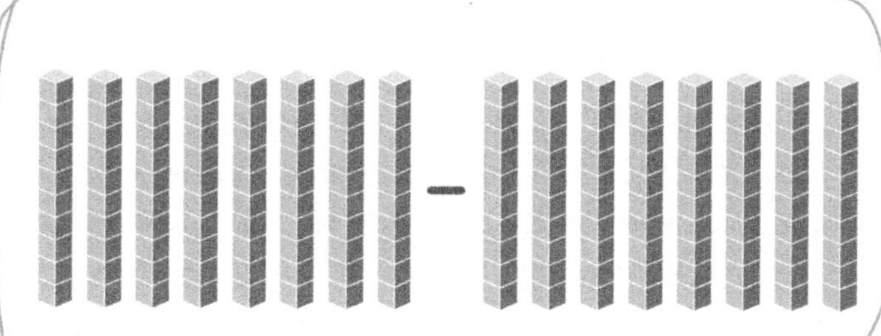

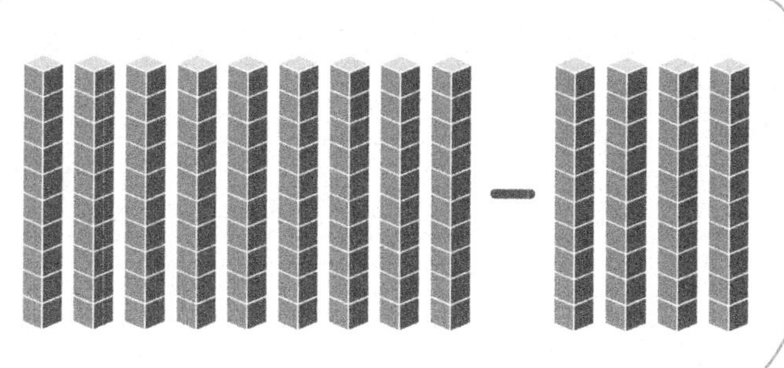

4. 90 - 40 = ☐

A. 20 B. 30 C. 40 D. 50

FITNESS PLANET

Let's get some fitness in! Go to page 167 to try some fitness activities.

FITNESS

Week 5 Writing Review

Topic 1 Temporal Words

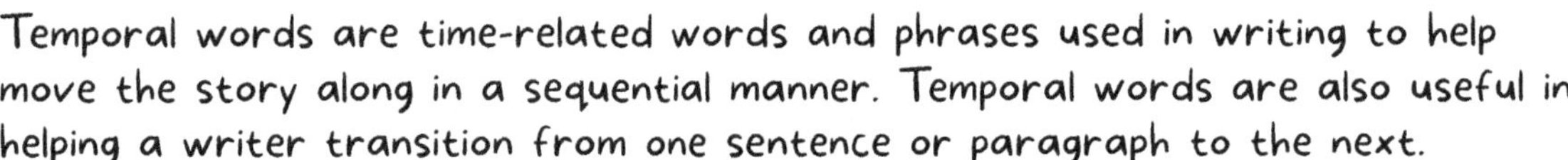

Temporal words are time-related words and phrases used in writing to help move the story along in a sequential manner. Temporal words are also useful in helping a writer transition from one sentence or paragraph to the next.

Examples of temporal words and phrases include:

in the beginning	**next**	**yesterday**
tomorrow	**last week**	**suddenly**
before	**during**	**in the end**
shortly after	**therefore**	**finally**

In the paragraph below, highlight or circle all the temporal words.

Last week, my dad and I went to the museum together. There were many different exhibits I wanted to see while we were there. First, we went to check out the dinosaurs. I love fossils! Suddenly, the lights dimmed, and we heard sounds like thunder. A "thunderstorm" was happening right in the exhibit. After it was done, we headed to the science exhibit. I got to build a boat to race down a river and observe turtles and fish in aquariums. Next, it was time to go to the theater to watch a short movie about astronauts. It was so interesting! Finally, the movie was over, and it was time to head home. What a great day I had with my dad!

Now it's your turn to practice writing sentences using temporal words. Write 3 sentences below. Underline the temporal word or phrase in each sentence.

1. ....

2. ....

3. ....

Now answer the following questions about the story you just read.

1. List the 3 exhibits that the child and his father visited while at the museum.

   A. ....

   B. ....

   C. ....

2. At the theater, they watched a short movie about:

   A. Leopards
   B. Space
   C. Dinosaurs
   D. Astronauts

3. What surprised the child about the dinosaur exhibit?

....

....

4. What types of live animals did the child and his dad see at the museum? Circle all that apply.

   A. Dinosaurs
   B. Turtles
   C. Fish
   D. Lizards
   E. Birds

5. Why did the child enjoy the dinosaur exhibit?

....

....

6. What did the child get to build while at the museum?

....

....

7. When did the child and his dad go to the museum?

   A. Today
   B. Last week
   C. Last month
   D. Yesterday

8. How did the child know a "thunderstorm" was happening?

....

....

Topic 1 Investigating Weather Hazards

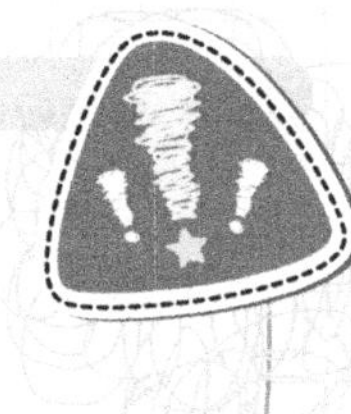

Weather hazards are extreme weather events that threaten people and/or property. Examples of weather hazards include: tornadoes, hurricanes, ice storms, high winds, blizzards and severe thunderstorms.

Using the information you researched last week on the climates of each of the seven regions in the United States, answer the following questions.

1. In what region might a tornado occur? ....

....

2. In what region might a hurricane occur? ....

....

3. In what region might a blizzard occur? ....

....

4. What types of weather hazards may occur in the West region? ....

....

5. What types of weather hazards may occur in the Mid-Atlantic region? ....

....

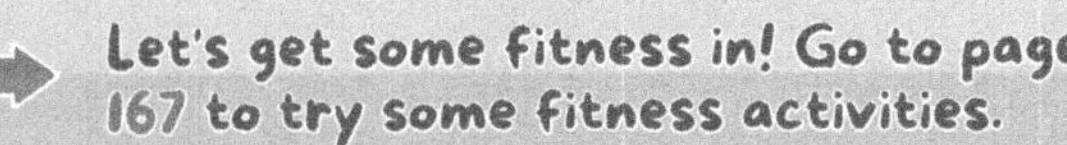

Topic 1 Investigating Weather Hazards

Now choose one weather-related hazard to think about more in-depth. What possible solutions can you think of that may reduce its impact on people and/or property?

For instance, in areas where hurricanes are common, buildings must meet be able to withstand winds of a certain speed to ensure they are safe during the high winds of a hurricane.

Weather Hazard: ....

Proposed Solution: ....

....

....

....

....

....

....

....

....

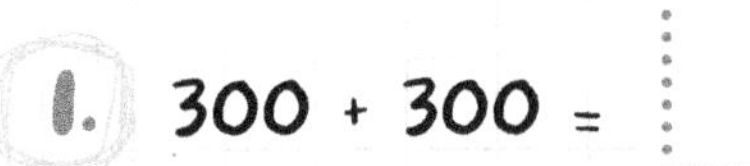

1. 300 + 300 =

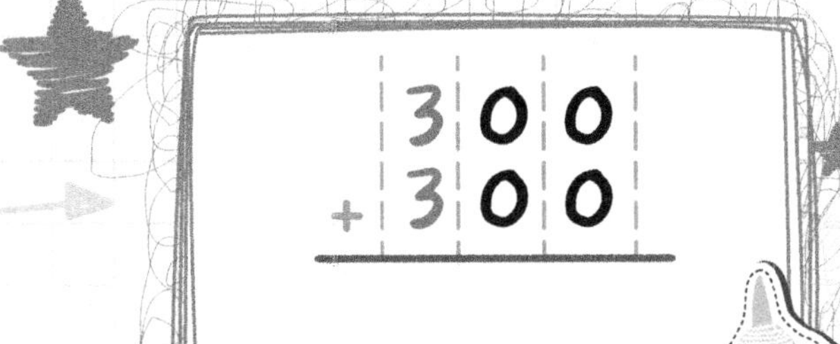

A. 400 B. 500 C. 600 D. 700

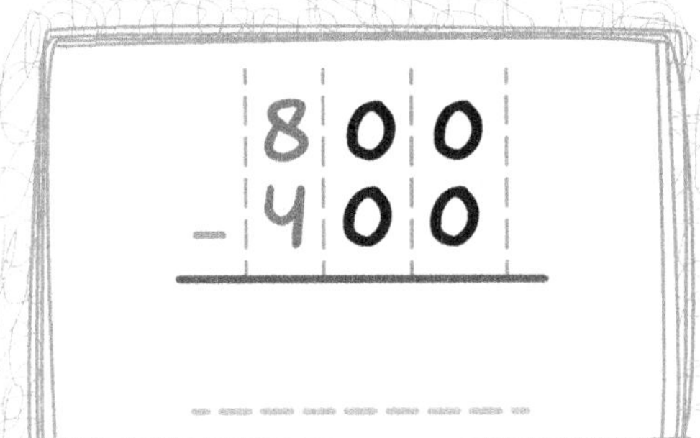

2. 800 - 400 =

A. 300 B. 400 C. 500 D. 600

3. 500 - 500 =

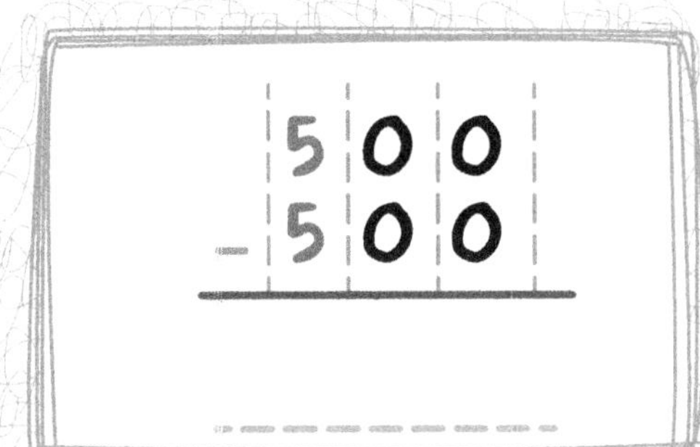

A. 300 B. 200 C. 100 D. 0

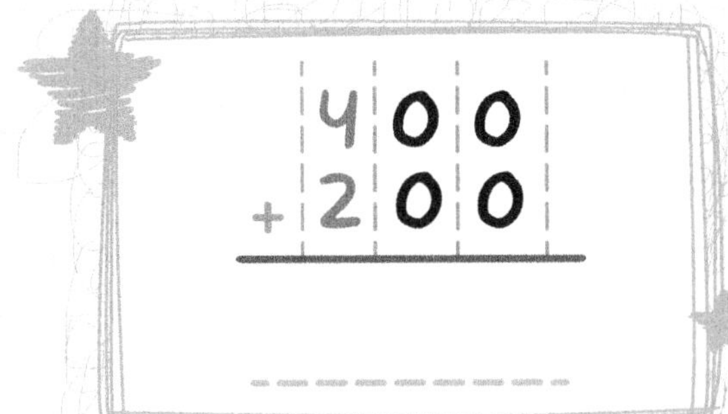

4. 400 + 200 =

A. 500 B. 600 C. 700 D. 800

5. 900 - 100 =

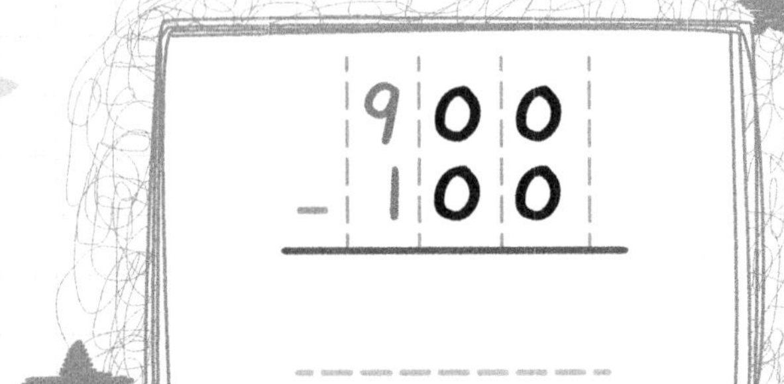

A. 800 B. 900 C. 1000 D. 1100

1. 700 - 200 = ☐

$$\begin{array}{r} 700 \\ -\ 200 \\ \hline \end{array}$$

A. 500 B. 400 C. 300 D. 200

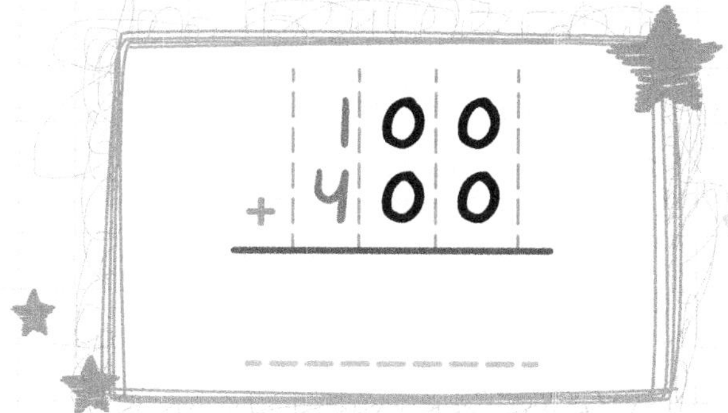

$$\begin{array}{r} 100 \\ +\ 400 \\ \hline \end{array}$$

2. 100 + 400 = ☐

A. 300 B. 400 C. 500 D. 600

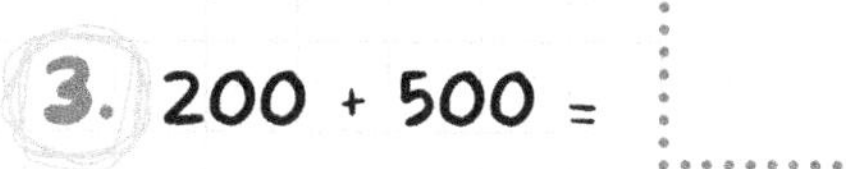

3. 200 + 500 = ☐

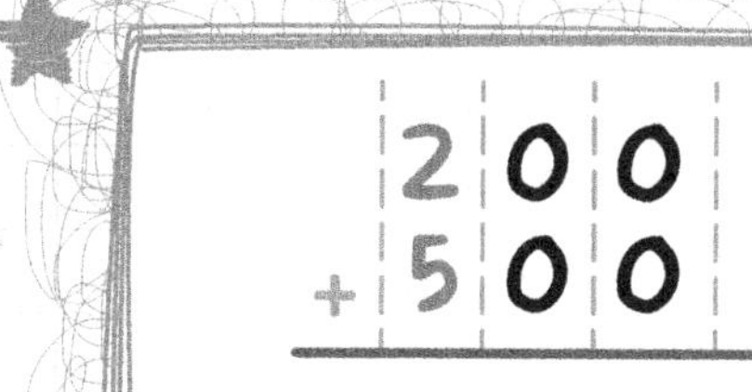

$$\begin{array}{r} 200 \\ +\ 500 \\ \hline \end{array}$$

A. 600 B. 700 C. 800 D. 900

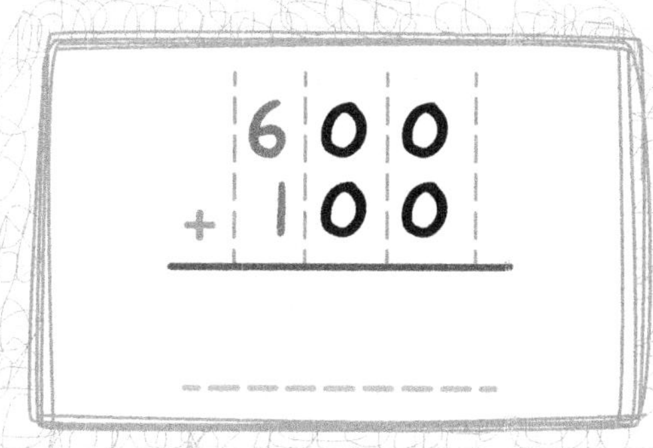

$$\begin{array}{r} 600 \\ +\ 100 \\ \hline \end{array}$$

4. 600 + 100 = ☐

A. 500 B. 600 C. 700 D. 800

5. 900 - 300 = ☐

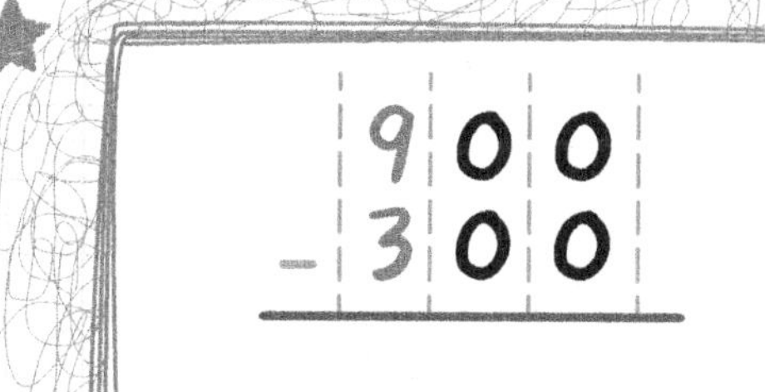

$$\begin{array}{r} 900 \\ -\ 300 \\ \hline \end{array}$$

A. 300 B. 400 C. 500 D. 600

Today, you will continue reviewing point of view by reading a passage and distinguishing your point of view from that of the author.

Today, my family took a trip to the zoo. While we were there, we attended a presentation about sloths. I learned so much about these amazing creatures!

Sloths are a type of mammal that live in Central and South America. Sloths mainly eat the buds, shoots, fruits and leaves of one kind of tree - the Cecropia tree. This tree is also the primary home to sloths. They only leave it, on average, once per week.

Sloths have long, sharp claws that help them climb trees. Despite having strong claws for climbing, these cute little animals are actually one of the slowest moving animals in the world. They only move 2 meters per minute! They also sleep a lot - about 10 hours per day.

After the presentation, I got to pet a cute sloth named Mo. He was so cuddly and friendly. As we were leaving the zoo, I told my parents that sloths are my new favorite animal!

Now answer the following questions about the passage.

1. What point of view is this passage being told from?

....

....

2. How do you know it is being told from that point of view?

....

....

3. If you were telling the story, what point of view would it be told from?

....

....

4. How is your point of view different from that of the child in the passage?

5. Which temporal words or phrases are used in the passage?

6. List two facts given in the passage.

7. List two opinions given in the passage.

On the lines below, rewrite the passage so that it is being told from the third person limited point of view.

FITNESS PLANET

Let's get some fitness in! Go to page 167 to try some fitness activities.

FITNESS

Topic 3 Mental Math: Adding and Subtracting by 10 and 100

1. 20 + 800 = ☐

A. 820 B. 1000 C. 280 D. 100

2. 300 - 30 = ☐

300
- 30

A. 250 B. 260 C. 270 D. 280

3. 400 + 90 = ☐

400
+ 90

A. 450 B. 460 C. 490 D. 500

4. 80 + 40 = ☐

80
+ 40

A. 100 B. 120 C. 140 D. 160

5. 660 - 20 = ☐

660
- 20

A. 640 B. 620 C. 600 D. 580

Today you are going to research a government leader from your community. This person may be part of the local, regional or state government. Answer the following questions about the government official you have chosen.

Government Leader: ....

Local, Regional or State Government: ....

Year Elected: ....

Key Issues Important to this Government Official:

1. ....
2. ....
3. ....
4. ....
5. ....

Key Accomplishments:

1. ....
2. ....
3. ....

Write a short paragraph about other things you learned about this person.

....

....

....

....

....

....

....

Grade 2-3

WEEK 6

Let's have a little fun with:

- measurement and data
- informative writing
- plant and animal traits
- geography exploration and more!

Week 6 Measurement and Data

Topic 1 Selecting and Using Appropriate Measurement Tools

1. What object should you use to measure a pencil?

A. Ruler
B. Yardstick
C. Measuring Tape

2. What object should you use to measure a cookie sheet?

A. Protractor
B. Yardstick
C. Measuring Tape

3. What object should you use to measure a textbook?

A. Ruler
B. Yardstick
C. Measuring Tape

4. What object should you use to measure the height of your house?

A. Ruler
B. Yardstick
C. Measuring Tape

5. What object should you use to measure a 2 pocket folder?

A. Ruler
B. Yardstick
C. Measuring Tape

FITNESS PLANET

Let's get some fitness in! Go to page 167 to try some fitness activities.

Week 6 Measurement and Data

Topic 1 Selecting and Using Appropriate Measurement Tools

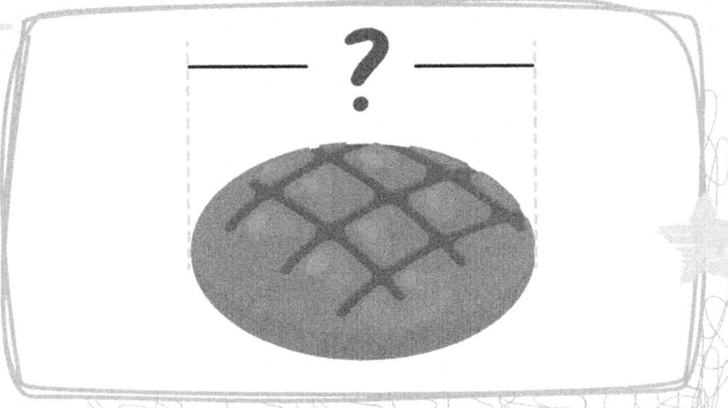

1. What object should you use to measure a loaf of bread?

A. Ruler
B. Yardstick
C. Measuring Tape

2. What object should you use to measure a stuffed bear?

A. Ruler
B. Yardstick
C. Measuring Tape

3. What object should you use to measure the length of a car?

A. Ruler
B. Yardstick
C. Measuring Tape

4. What object should you use to measure the height of a door?

A. Ruler
B. Yardstick
C. Protractor

Week 6 Writing Review

Topic 1 Informative/Explanatory Writing

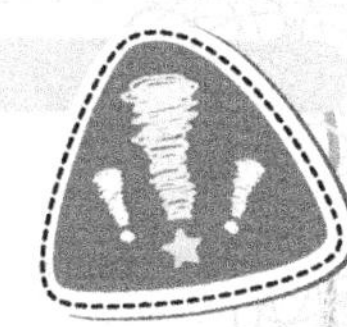

Informative writing is writing that teaches the reader about a specific topic.

Today you will practice writing a short informative piece. This organized paragraph on a specific topic will include an introduction of the topic, facts and other details that develop the topic and a concluding statement. You will begin by mapping out your writing first and then put it together into a well-organized paragraph.

Think about a real thing that you know a lot about. Examples might include: a plant, animal, game, sport, geographic location or subject area. This will be your topic for this writing piece. A **topic** is what you are writing about. You will always start any type of writing with a topic and often use it to introduce or begin your piece of writing.

Topic: ....

What is your topic? It should be explicitly stated at the beginning of your writing piece in a well-developed sentence.

Supporting Details: ....

Think of at least 3 facts or details about your topic. You will use these as supporting details in your writing. You should develop each fact into a sentence or two in your writing piece.

Detail #1: ....

Detail #2: ....

Detail #3: ....

A **concluding statement** is a way to wrap up your writing. It should sum up your paragraph and give the reader an idea of how you feel about the topic.

Concluding statement: ....

Now you will use the graphic organizer you created above to write a well-organized informative piece about your topic on the lines below. Be sure your writing includes an introduction, at least 3 facts or details to develop your topic and a concluding statement. Remember to write in complete sentences and use capital letters and correct punctuation.

Week 6 Language Review

Topic 2 Sentence Types

There are 3 different types of sentences. They are **simple**, **compound** and **complex.** Great writers use different sentence types in their writing to keep the reader interested and engaged.

Examples:

Simple	I read a book.
Compound	I read a book, and I learned something new about dolphins.
Complex	Even though I read a book about dolphins, I still don't understand echolocation.

Practice by identifying the sentence type for each example below.

1. Because he was sick with the flu, I did not see Joe at school today.
   A. Simple
   B. Compound
   C. Complex

2. I played a game with my sister.
   A. Simple
   B. Compound
   C. Complex

3. We played soccer after school, and I made a goal.
   A. Simple
   B. Compound
   C. Complex

4. My best friend is funny.
   A. Simple
   B. Compound
   C. Complex

5. When Mrs. Smith is absent, the students act silly all day long.
   A. Simple
   B. Compound
   C. Complex

Topic 1 Plant and Animal Traits

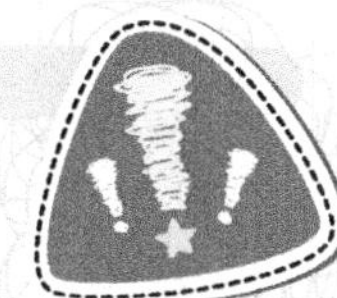

Plants and animals have **traits,** or characteristics, that are **inherited,** or passed down from their parents. These traits are what make plants and animals look similar to their parents and other family members.

Examples of Inherited Traits

Plants	Animals
Flower color	Eye color
Flower shape	Hair color and type
Leaf shape	Height
Plant height	Bone structure
Seed shape	Skin color

FITNESS PLANET → Let's get some fitness in! Go to page 167 to try some fitness activities.

FITNESS

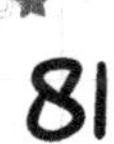

Now you will examine two plants of the same species and compare and contrast them.

Materials Needed

- 2 plants or flowers of the same species
- Magnifying glass

Procedure

1. Examine the two plants or flowers, looking specifically for ways they are similar and different.
2. Answer the questions below.

Follow-Up Questions

1. List 3 ways the plants/flowers are alike.

....

....

2. List 3 ways the plants/flowers are different.

....

....

3. Why do you think they are similar?

....

....

Topic 2 Estimate lengths

1. What is a reasonable estimate for the length of a pencil?

A. 9 inches
B. 90 inches
C. 19 inches
D. 900 inches

2. What is a reasonable estimate for the length of a cookie sheet?

A. 8 inches
B. 18 inches
C. 88 inches
D. 800 inches

3. What is a reasonable estimate for the length of the height of your house?

A. 22 feet
B. 22 inches
C. 22 centimeters
D. 22 meters

4. What is a reasonable estimate for the height of a tree?

A. 18 feet
B. 1 inch
C. 80 inches
D. 1 centimeter

Week 6 Measuring an Object

Topic 3 Measure each line using centimeters

1. Measure the line below.

....

2. Measure the line below.

....

3. Measure the line below.

....

4. Measure the line below.

....

5. Measure the line below.

....

6. Measure the line below.

....

7. Measure the line below.

....

8. Measure the line below.

....

Read the passage below.

It was a dark and stormy night. Georgia was feeling scared as she lay in her bed, trying hard to fall asleep. The tree outside her window was swaying in the wind and thunder boomed every couple of minutes. The streetlight outside cast shadows on her wall that made her feel like there was something creeping around her room. She pulled the covers up over her head and squeezed her eyes closed. She really wanted her mom to come back into her room, but she knew she was trying to get her baby brother down to sleep.

Suddenly, she heard a creaking sound, and she realized her bedroom door was slowly opening. Someone was entering her bedroom. Just as she was about to scream for her mom, the light flipped on and she saw her dad standing in the doorway.

"I thought you might like some company," he said, crossing the room to her bed.

"Oh dad, thank goodness. I was so scared," Georgia replied.

As her dad climbed onto the top bunk above her, Georgia relaxed into her bed and knew everything would be just fine.

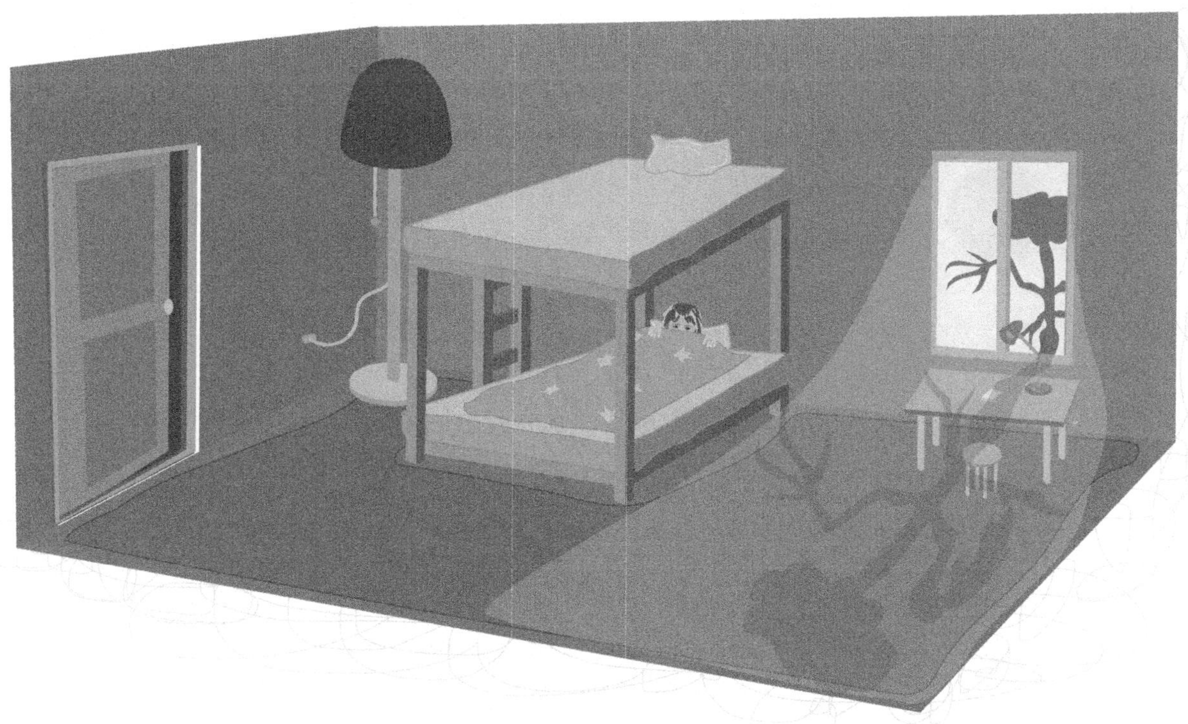

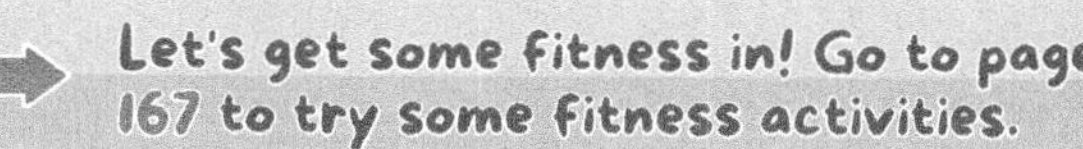

Week 6 Reading Passage

Topic 3 Literature

Answer the questions below about the passage.

1. How was Georgia feeling at the beginning of the story?

2. Highlight the text that helps you to know how she was feeling at the beginning of the passage.

3. How was Georgia feeling at the end of the story?

4. What words does the author use to help you know how Georgia is feeling throughout the story?

5. What is causing Georgia to feel scared in the story?

6. How might the story change, if Georgia's dad did not come into her bedroom?

Week 6 Measuring an Object

Topic 3 Measure each line using inches

1. Measure the line below. ....

2. Measure the line below. ....

3. Measure the line below. ....

4. Measure the line below. ....

5. Measure the line below. ....

6. Measure the line below. ....

7. Measure the line below. ....

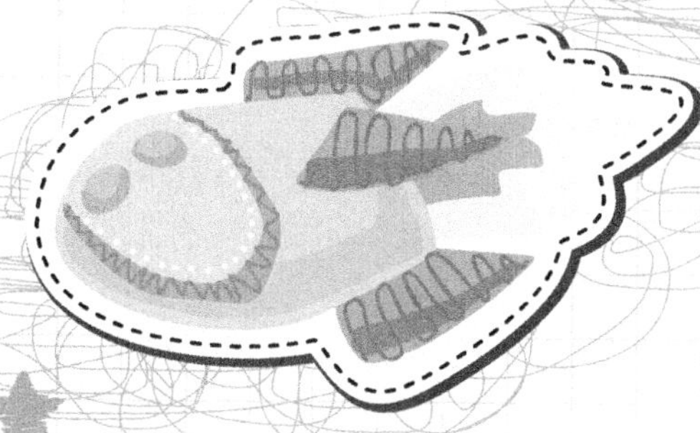

Week 6 Social Studies

Topic 1 Geography Exploration

Today you will explore the geography of the region in which you live (Northwest, West, Southwest, Midwest, Southeast, Mid-Atlantic, Northeast) by labeling different features on a map. To complete this activity, you will need to print out a blank map of your region.

Key Vocabulary

Physical features include bodies of water and landforms such as oceans, lakes, mountains, rivers, volcanoes, peninsulas, etc.

Political features include boundaries for countries and states and locations of cities.

Labels help to identify different features on a map.

Map symbols are used to mark certain features or locations. They are defined in the **map key or legend** which aids the reader in understanding the map.

Now you will apply these vocabulary terms by labeling your own map. Follow the directions below.

On the map of your region, complete the following tasks:

1. Region you are exploring:
2. Draw boundary lines for each state included in the region
3. Mark any large cities and state capitals located in the region
4. Draw and/or label any bodies of water found in the region
5. Draw and/or label any landforms found in the region
6. Create a map key or legend to help readers understand any symbols you used

Grade 2-3

WEEK 7

Let's learn about:

- spelling
- plants and basic needs
- number lines
- telling time and more!

Circle the smaller objects.

1. 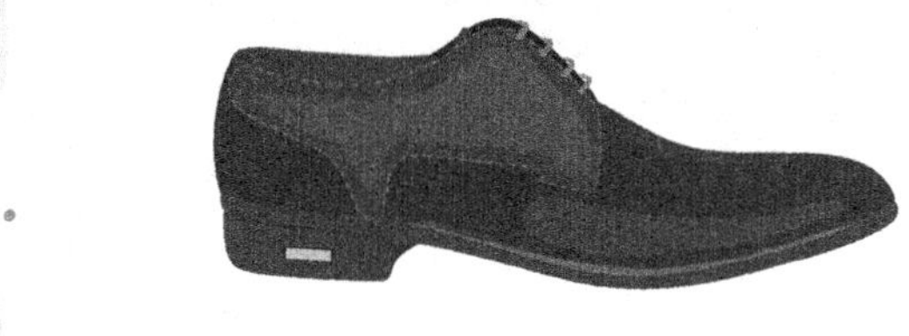 

2. 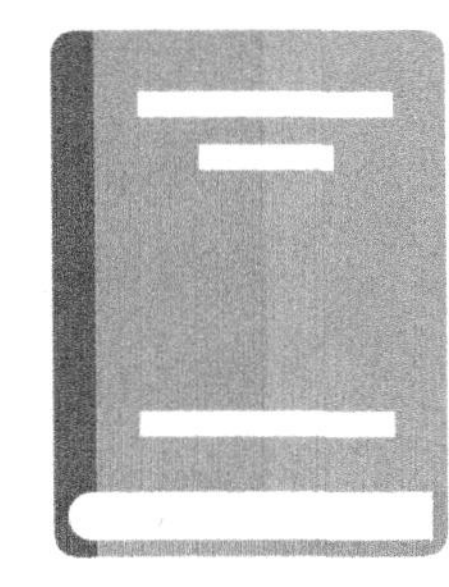 

Circle the larger objects.

3. 

4.  

Week 7 Measurement and Data

Topic 1 Looking at Length

1. What should you use to measure an apple?

....

A. Inch C. Yard
B. Foot D. Mile

2. What should you use to measure a chalkboard?

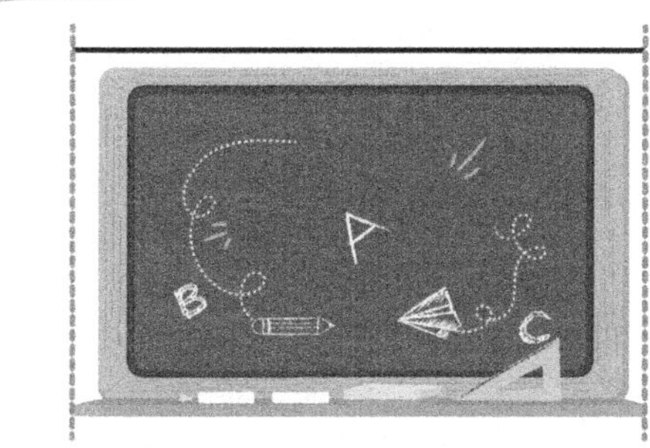

....

A. Centimeter C. Kilometer
B. Meter D. Millimeter

3. What should you use to measure a bookshelf?

....

A. Inch C. Yard
B. Foot D. Mile

4. What should you use to measure a TV?

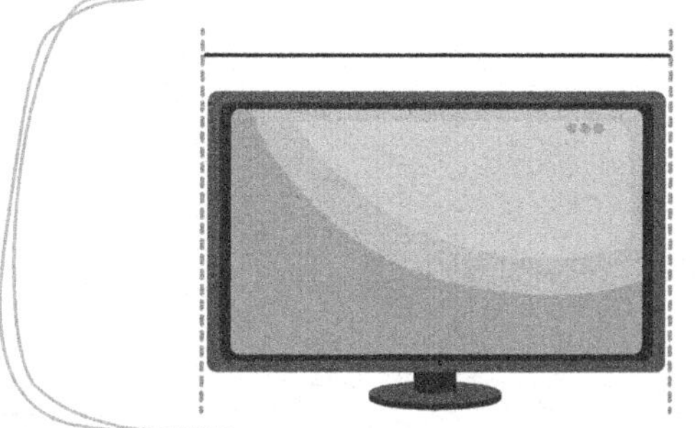

....

A. Inch C. Yard
B. Foot D. Mile

5. What should you use to measure medicine in a cup?

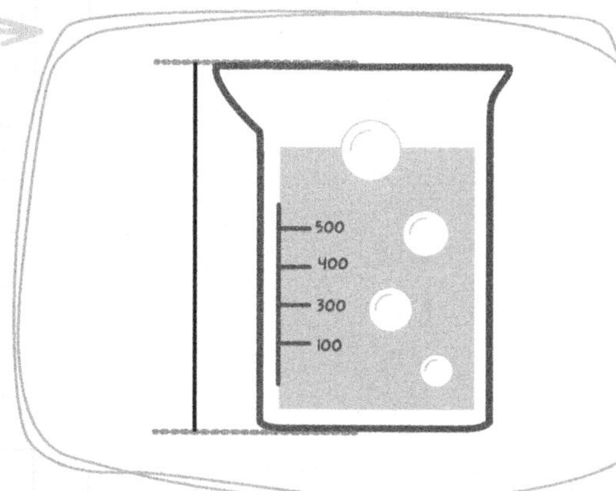

....

A. Centimeter C. Kilometer
B. Meter D. Milliliter

Week 7 Language Review

Topic 1 Spelling

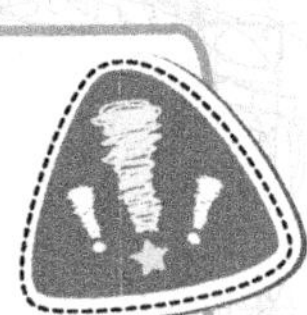

High frequency words are words that are used frequently in books.

Sight words are words that do not follow conventional spelling rules and must be memorized.

Because both high frequency words and sight words are very common in reading and writing, it is important to be able to read them, as well as spell them.

Choose the correct spelling of the following high frequency and sight words.

1.
A. Abot C. Abowt
B. Abbout D. About

2.
A. Done C. Dune
B. Dun D. Don

3.
A. Laf C. Laugh
B. Laph D. Lagh

4.
A. Clean C. Cleen
B. Klean D. Kleen

5.
A. Lite C. Liet
B. Light D. Lit

Topic 1 Spelling

Now you will practice adding suffixes to base words. Read the directions, then rewrite the word on the line. Be careful to spell the new word correctly.

1. Sit + ing = ...
2. Happy + ness = ...
3. Child + hood = ...
4. Shred + ed = ...
5. Knife + es = ...
6. Smile + ed = ...
7. Beauty + ful = ...

Now you will practice changing singular nouns to plural nouns. Some may be irregular so be careful to spell the new word correctly.

1. Baby = ...
2. Leaf = ...
3. Bush = ...
4. Tooth = ...
5. Fish = ...
6. Lady = ...
7. Child = ...
8. Goose = ...

Week 7 Science

Topic 1 The Basic Needs of Plants

Today, you will explore the basic needs of plants. **Basic needs** are the things plants need to grow. These include: water, air, sunlight and nutrients. You are going to conduct an experiment to observe the differences in the growth and development of plants when they have access to some basic needs but not others.

Materials Needed:

* Seeds (bean or flower seeds work best)
* 3 small cups
* Soil

Procedure:

1. Fill each of the cups halfway with soil, push 1 seed into the soil and cover it with a bit more soil.

2. Label the cups as Plant 1, Plant 2 and Plant 3.

3. Decide which basic needs you will give to each plant and record in the first column below (i.e. you may withhold water from Plant 1, sunlight from Plant 2 and both water and sunlight from Plant 3).

4. Observe each plant over time and record your observations below. You may choose to wait until the seed begins to sprout to record your observations.

5. Be sure to observe the soil, stem, leaves and flowers.

Topic 1 The Basic Needs of Plants

Basic Needs Withheld from Plant 1: ....

Basic Needs Withheld from Plant 2: ....

Basic Needs Withheld from Plant 3: ....

	Plant 1	Plant 2	Plant 3
Day 1			
Day 2			
Day 3			
Day 4			
Day 5			
Day 6			
Day 7			

FITNESS PLANET → Let's get some fitness in! Go to page 167 to try some fitness activities.

Topic 2 Whole Numbers on a Number Line

1. What whole number would come after 11 on the number line?

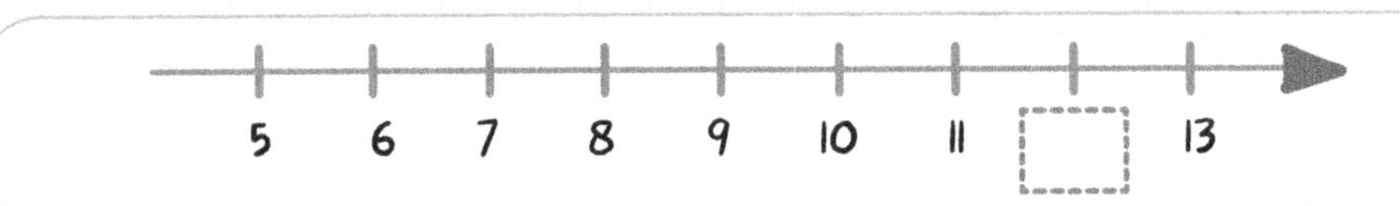

A. 12 B. 13 C. 14 D. 15

2. What whole number would come after 47 on the number line?

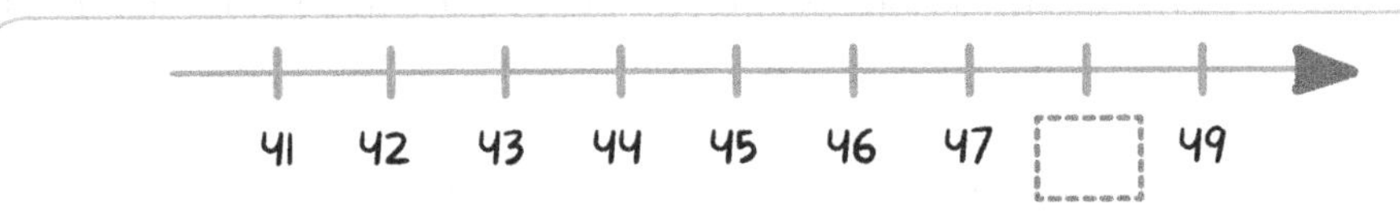

A. 46 B. 45 C. 48 D. 49

3. What whole number would come after 52 on the number line?

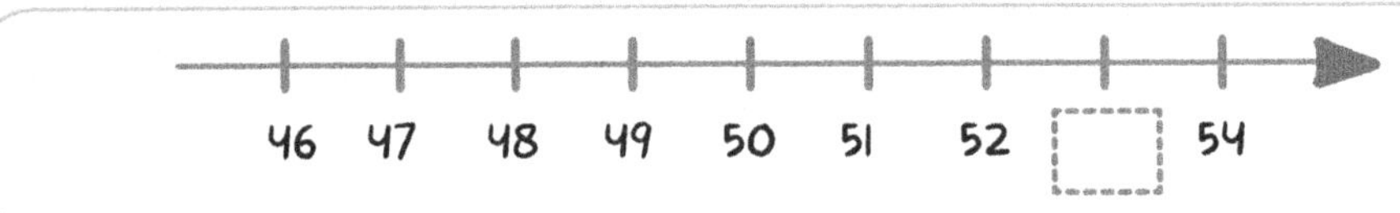

A. 50 B. 51 C. 55 D. 53

4. What whole number would come after 96 on the number line?

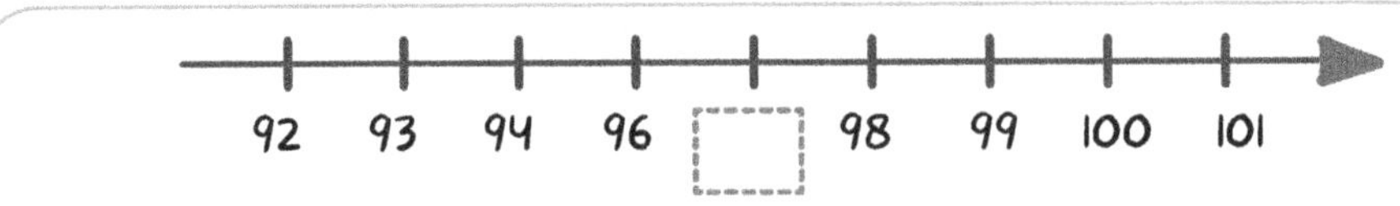

A. 95 B. 97 C. 99 D. 101

5. What whole number would come after 104 on the number line?

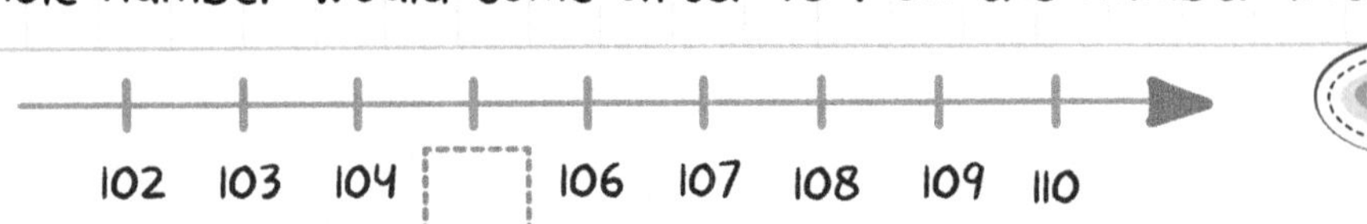

A. 105 B. 106 C. 107 D. 108

How would the number look on a digital clock?

1. six fourteen

A. 4:16 C. 1:14
B. 6:14 D. 6:41

2. two twenty nine

A. 2:29 C. 2:92
B. 9:22 D. 2:29

3. one forty five

A. 1:54 C. 5:41
B. 4:15 D. 1:45

4. nine thirty six

A. 6:39 C. 3:29
B. 9:36 D. 3:06

5. four twenty one

A. 4:12 C. 4:21
B. 2:14 D. 1:24

6. twelve o two

....

7. seven fifty one

....

Read the following passage. Then, answer the questions that follow.

Keira was playing with friends at the playground near her house after school. She was having a great time swinging high on the swings. Keira was so excited that it was Friday and she had two days off from school.

Her friend, Michael, ran over and challenged Keira to a race on the monkey bars. Joey was timing each kid to see how fast they could get across. Keira had been practicing so she ran over, excited to compete with her friends. She waited her turn, cheering on the others as they crossed. As each of her friends finished, Joey yelled out a time, and all the kids cheered or booed, depending on how fast they had gone.

Finally, it was Keira's turn. She felt a little nervous with all the other kids watching her. Joey started the stopwatch, and off she went! She was halfway across when suddenly she lost her grip and tumbled to the ground. She lay on the dirty mulch, her arm throbbing in pain. Her friends all ran over, and Keira felt embarrassed, as she began to cry.

Michael ran across the street to get Keira's mom. They came running over, and Keira felt comforted just seeing her mom there. Her mom scooped her up and carried her back home, straight up to her bedroom. She lay on her bed crying, as her mom went to the kitchen to get a bag of ice for her arm. When she returned, her mom laid down on the bed next to her, holding the ice on her arm. Keira was so glad to have her mom next to her, taking care of her. She knew everything was going to be ok.

On the chart below, list all the different feelings Keira experiences throughout the story. You can use words from the passage or put the answers into your own words.

Answer the following questions about the story you just read.

1. What motivated Keira to compete on the monkey bars with her friends?

2. Why did some kids get cheered and some kids get booed after their turn on the monkey bars?

3. What does the phrase "scooped her up" mean?

4. Which event happens first in the story?
   A. Keira waits her turn on the monkey bars.
   B. Her mom lays her down on her bed.
   C. She begins going across the monkey bars.
   D. Keira was swinging.

5. What action causes Michael to run to get Keira's mom?

6. Why do you think Keira is embarrassed when she falls off the monkey bars?
   A. The kids laugh at her.
   B. She gets dirt all over her clothes.
   C. She is crying.
   D. Her mom comes to the playground.

7. What causes Keira to run over to the monkey bars?

8. Why do you think Keira felt nervous when it was her turn on the monkey bars?

Let's get some fitness in! Go to page 167 to try some fitness activities.

Week 7 Measurement and Data

Topic 3 Telling Time

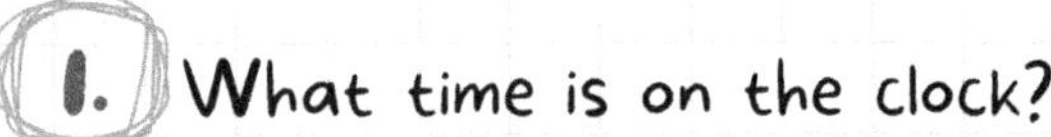

1. What time is on the clock?

....

....

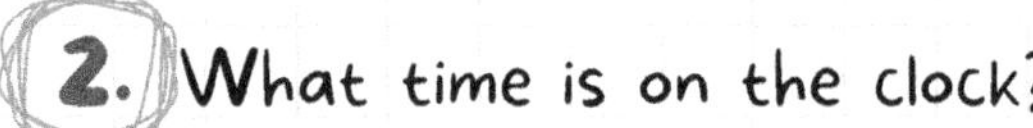

2. What time is on the clock?

....

....

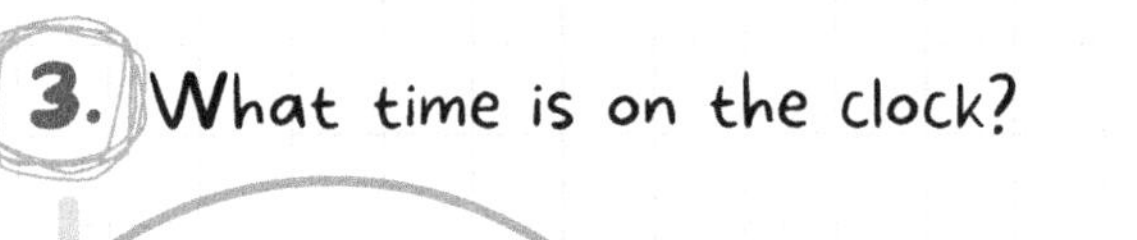

3. What time is on the clock?

....

....

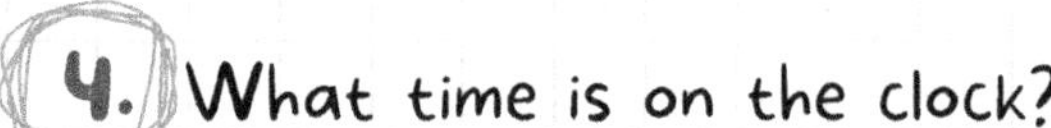

4. What time is on the clock?

....

....

5. What time is on the clock?

....

....

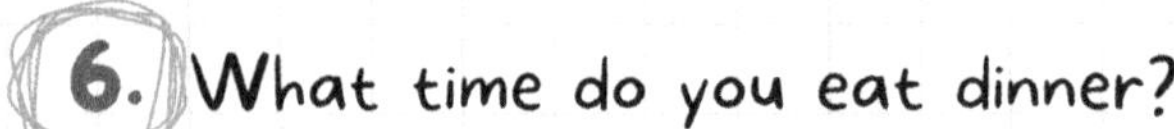

6. What time do you eat dinner?

....

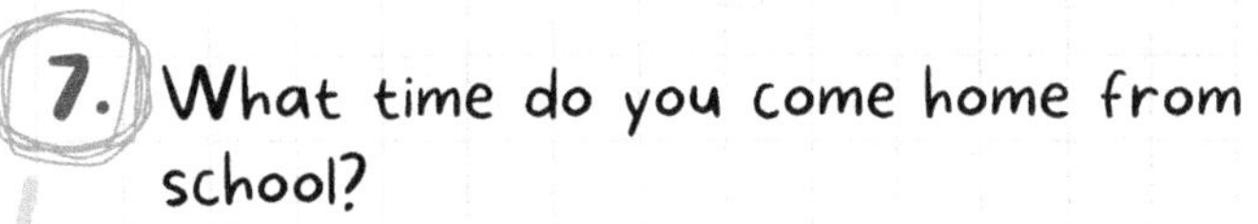

7. What time do you come home from school?

....

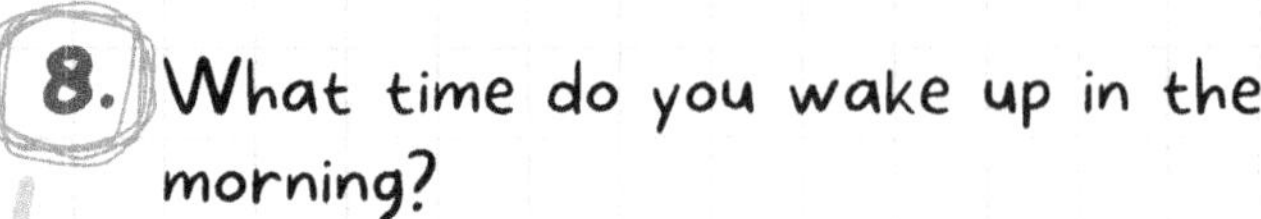

8. What time do you wake up in the morning?

....

Week 7 Social Studies

Topic 1 Geography - Spatial Terms

Key Vocabulary

1. **Hemisphere** - half of the Earth, as divided by the Equator or an imaginary line that passes through the Poles

2. **Cardinal directions** - the four main directions: North, South, East and West

3. **Intermediate directions** - the directions that lie in between each of the four cardinal directions: northeast, southeast, southwest, northwest

4. **Compass rose** - a circle, found on a map or globe, that identifies the cardinal directions

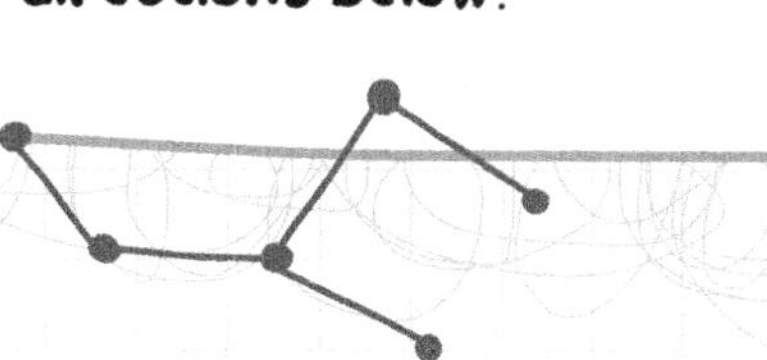

Today, you will explore these spatial terms using a map of the United States. Follow the directions below.

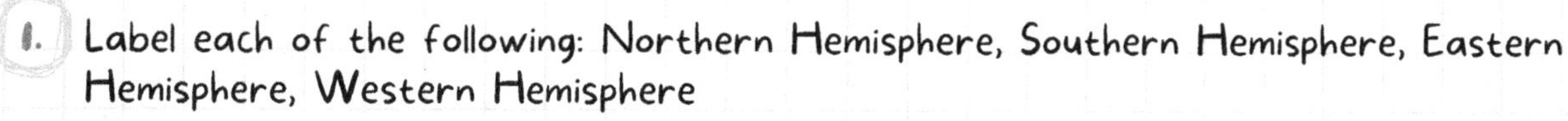

1. Label each of the following: Northern Hemisphere, Southern Hemisphere, Eastern Hemisphere, Western Hemisphere

2. Draw a compass rose on your map and label the cardinal and intermediate directions

3. If you were traveling from Europe to Asia, what direction would you travel?

4. If you were traveling from Africa to Antarctica, what direction would you travel?

5. Where might you be traveling, if you begin in South America and travel north?

6. What hemisphere is Australia located in?

7. Where might you be traveling, if you begin in Africa and travel northeast?

8. What hemisphere is Europe located in?

Grade 2-3

WEEK 8

It's time to learn about:

* money math
* narrative writing
* graphs
* commas in letters

Week 8 Measurement and Data

Topic 1 Word Problems Involving Money

1. How much is two pennies and three quarters?

A. $0.52
B. $0.77
C. $0.28
D. $0.32

2. How much is one dime, one nickel and two quarters?

A. $0.45
B. $0.77
C. $0.65
D. $0.17

3. How much is two pennies and three dimes?

A. $0.32
B. $0.42
C. $0.52
D. $0.62

4. How much is one penny, one nickel, one dime, one quarter?

A. $0.40
B. $0.41
C. $0.39
D. $0.38

5. How much is three dimes and one penny?

A. $0.11
B. $0.21
C. $0.31
D. $0.41

Topic 1 Word Problems Involving Money

1. What combination of coins would give you $1.26?

2. What combination of coins would give you $1.08?

3. What combination of coins would give you $0.72?

4. What combination of coins would give you $1.20?

5. What combination of coins would give you $0.16?

6. What combination of coins would give you $0.71?

7. What combination of coins would give you $1.33?

8. What combination of coins would give you $0.04?

Week 8 Writing Review

Topic 1 Narrative Writing

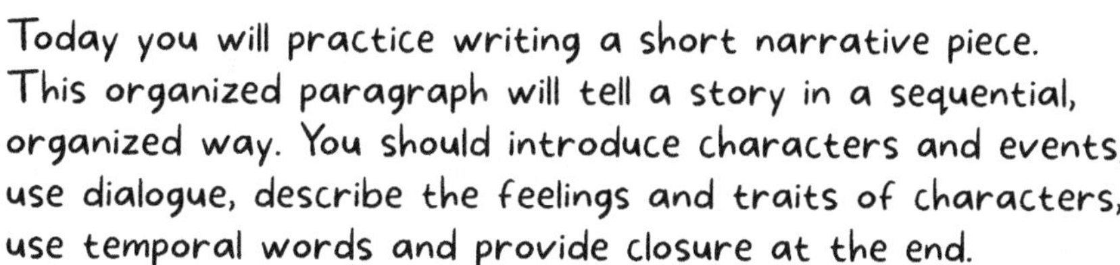

Narrative writing is writing that tells a story.

Today you will practice writing a short narrative piece. This organized paragraph will tell a story in a sequential, organized way. You should introduce characters and events, use dialogue, describe the feelings and traits of characters, use temporal words and provide closure at the end.

You will begin by mapping out your writing first and then put it together into a well-organized paragraph with a beginning, middle and end. Think about a story you want to tell. It can be real or imaginary.

Beginning: Establish the situation and introduce your characters.

Middle: Develop the story in a sequential manner.

Ending: Provide a sense of closure.

Now you will use the graphic organizer you created above to write a well-organized narrative piece on the lines below. Be sure your writing includes a beginning, middle and end. Remember to write in complete sentences and use capital letters and correct punctuation.

Week 8 Science

Topic 1 A Closer Look at Animals

Some animals live in groups that help its members survive. Animals may form groups for a variety of reasons. These may include:

1. **For Protection**
2. **To Hunt for Food**
3. **To Care for Young**
4. **To Build Shelters**

Examples of animals that live or travel in groups include: birds, fish, wolves, monkeys and beavers.

Answer the following questions.

1. How might living in a group help keep animals protected?

....

....

2. Give an example of an animal that lives in a group for protection.

....

....

3. How might living in a group help animals look for food?

....

....

4. Give an example of an animal that hunts in a group.

....

....

5. How might living in a group help animals to care for their young?

....

....

FITNESS PLANET

Let's get some fitness in! Go to page 167 to try some fitness activities.

6. Give an example of an animal that lives in a group to care for their young.

....

....

7. How might living in a group help animals to build shelters?

....

....

8. Give an example of an animal that lives in a group to build shelters.

....

....

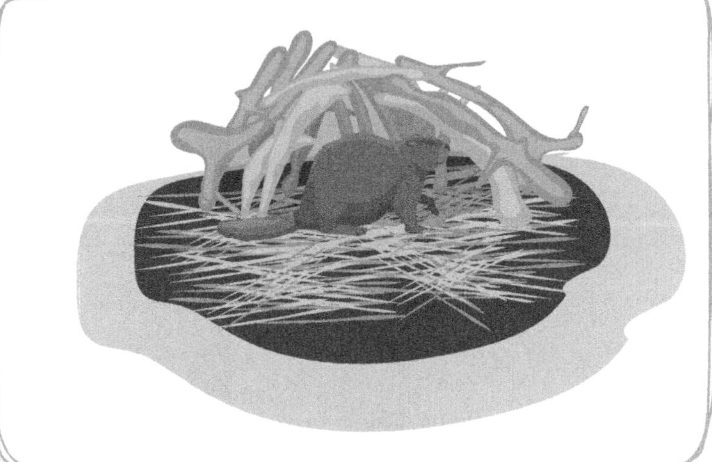

Topic 2 Generating Measurement Data

Find an object around you to measure. Write down the name of the object and then write an exact length in inches using your ruler.

1. Object: .... Measurement: ....

2. Object: .... Measurement: ....

3. Object: .... Measurement: ....

4. Object: .... Measurement: ....

5. Object: .... Measurement: ....

6. Object: .... Measurement: ....

7. Object: .... Measurement: ....

8. Object: .... Measurement: ....

9. Object: .... Measurement: ....

Week 8 Measurement and Data

Topic 3 Representing Data in Picture and Bar Graphs

Use the graph below to answer the following questions

1. How many students like dogs?

A. 9 B. 10 C. 11 D. 12 ....

2. How many students like cats?

A. 7 B. 8 C. 9 D. 10 ....

3. How many students were surveyed in all?

A. 19 B. 20 C. 21 D. 22 ....

4. How many more students like dogs than cats?

A. 3 B. 4 C. 5 D. 6 ....

5. If three more students like cats, how many students would like cats total?

A. 9 B. 10 C. 11 D. 12 ....

Week 8 Language Review

Topic 2 Commas in Addresses

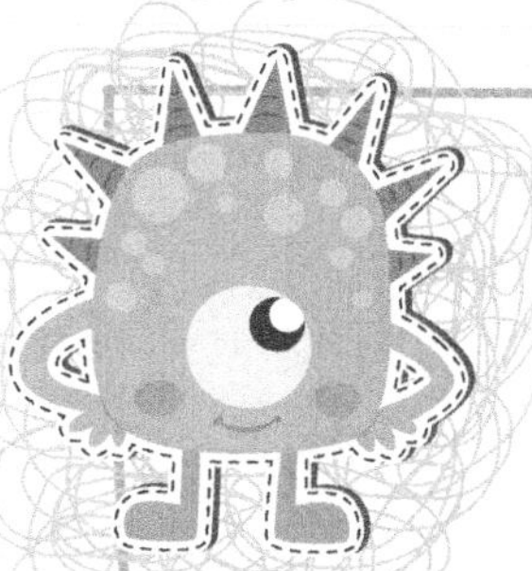

Commas are used often in writing. When writing addresses, there are 3 rules that must be followed for comma usage.

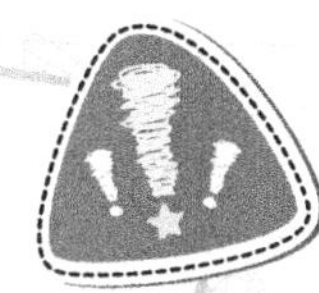

1. **When addressing an envelope, use a comma between the city and state.**
   Example: 123 Sycamore Street,
   West Bend, Oregon 12345
2. **When writing out an address, use a comma after the street and between the city and state.**
   Example: 123 Sycamore Street, West Bend, Oregon 12345
3. **When using an address in a sentence, use a comma after the street, between the city and state and after the entire address.**
   Example: I lived at 123 Sycamore Street, West Bend, Oregon, for six years.

Now it's your turn to practice.

1. Which address is written correctly?

   A. 538 Elm Street, Smith, Minnesota, 47659
   B. 538 Elm Street Smith, Minnesota 47659
   C. 538 Elm Street, Smith, Minnesota 47659
   D. 538 Elm Street, Smith Minnesota 47659

2. Which address is written incorrectly?

   A. 701 Bugle Lane, Park, Idaho 46544
   B. 2708 Main Street Elmhurst, Illinois, 60807
   C. 3967 Central Avenue, Tampa, Florida 30809
   D. 451 Carr Boulevard, Rush, California 98177

3. Correctly write the incorrect address from #2 on the lines below.

....

....

....

....

FITNESS PLANET

Let's get some fitness in! Go to page 167 to try some fitness activities.

Week 8 Reading Passage

Topic 2 Informational Text

How to Make a Peanut Butter and Jelly Sandwich

First, you need to gather all the ingredients needed for a peanut butter and jelly sandwich. This includes bread, peanut butter, jelly, a knife and a plate. Next, lay each piece of bread on the plate. Then, use the knife to spread the peanut butter on one of the pieces of bread and jelly on the other piece of bread. After that, place one piece of bread on top of the other. Lastly, enjoy your peanut butter and jelly sandwich!

Now you are going to answer some questions about the passage.

1. What is the logical connection between each of the sentences in the passage?
   - A. Comparison
   - B. Cause and effect
   - C. Sequential

2. How do you know this is the connection between the sentences?

3. What types of words help distinguish each step in the process?

4. Highlight all the transition words in the passage.

5. List 3 other examples of transition words.

6. How are transition words helpful in writing?

Week 8 Measurement and Data

Topic 3 Representing Data in Picture and Bar Graphs

1. How many students like winter?

   A. 5 B. 6 C. 7 D. 8

2. How many students like spring?

   A. 3 B. 4 C. 5 D. 6

3. What is the season the most amount of students like?

   A. spring C. fall
   B. summer D. winter

4. What is the season the least amount of students like?

   A. spring C. fall
   B. summer D. winter

5. How many students were surveyed?

   A. 32 B. 33 C. 34 D. 35

6. How many more students like summer over spring?

   A. 9 B. 10 C. 11 D. 12

Week 8 Social Studies

Topic 1 Regions

Regions are areas that share similar physical and cultural characteristics.

Physical characteristics include bodies of water and landforms such as mountains, hills, valleys, volcanoes, etc.

Cultural characteristics include population, transportation networks and religion and customs.

Let's explore the region you live in. Answer the questions below.

1. What state do you live in? ....

....

....

2. What region is your state in? ....

....

....

3. Describe the physical characteristics of the region you live in. ....

....

....

4. Describe the cultural characteristics of the region you live in. ....

....

....

5. How are the physical characteristics of your state similar to the physical characteristics of a neighboring state?

....

....

6. How are the physical characteristics of your state different from the physical characteristics of a neighboring state?

....

....

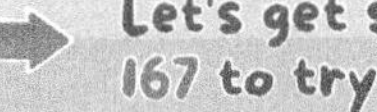

Let's get some fitness in! Go to page 167 to try some fitness activities.

Grade 2-3

WEEK 9

Get ready to explore:

- shapes and attributes
- verb tenses
- properties of sound
- climate regions and more!

Week 9 Geometry

Topic 1 Attributes of Shapes

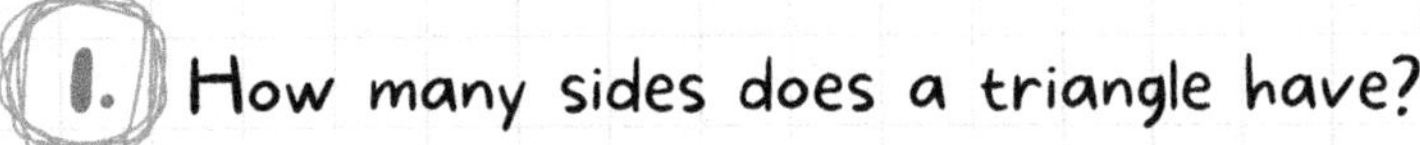

1. How many sides does a triangle have?

A. 3 B. 4 C. 5 D. 6

2. How many sides does a quadrilateral have?

A. 6 B. 5 C. 4 D. 3

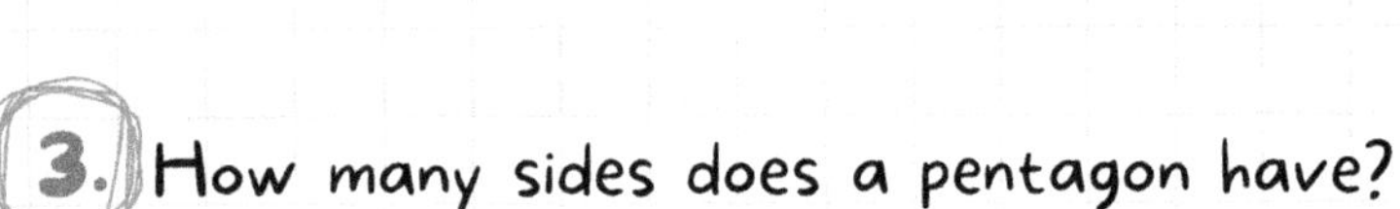

3. How many sides does a pentagon have?

A. 6 B. 5 C. 4 D. 3

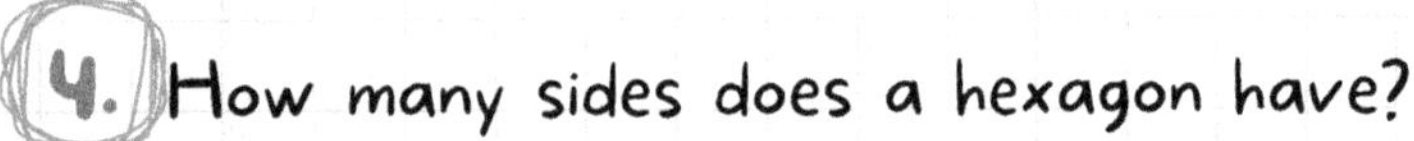

4. How many sides does a hexagon have?

A. 3 B. 4 C. 5 D. 6

5. Which shape has 3 angles?

A. Quadrilateral C. Hexagon
B. Pentagon D. Triangle

6. Which shape has 4 angles?

A. Pentagon C. Hexagon
B. Triangle D. Quadrilateral

7. Which shape has 5 angles?

A. Quadrilateral C. Hexagon
B. Pentagon D. Triangle

Week 9 Geometry

Topic 1 Attributes of Shapes

1. Draw a triangle.

2. Draw a quadrilateral.

3. Draw a pentagon.

4. Draw a hexagon.

5. Draw a cube.

6. Draw a triangle.

Week 9 Language Review

Topic 1 Verb Tenses

Verb tenses are used to indicate whether something happened in the past, present or future.

1. **Past Tense: the action already happened**
   Example: The plane landed twenty minutes ago.
2. **Present Tense: the action is happening right now**
   Example: The plane is landing right now.
3. **Future Tense: the action will happen in the future**
   Example: The plane will land in fifteen minutes.

Now you will practice identifying and using correct verb tense in sentences.

1. We .... last Tuesday.
   A. rode horses
   B. are riding horses
   C. will be riding horses

2. Sam .... a really good book in class right now.
   A. read
   B. is reading
   C. will read

3. The builders will construct a new skyscraper next year.
   A. Past tense
   B. Present tense
   C. Future tense

4. Julie is walking her dog to the park.
   A. Past tense
   B. Present tense
   C. Future tense

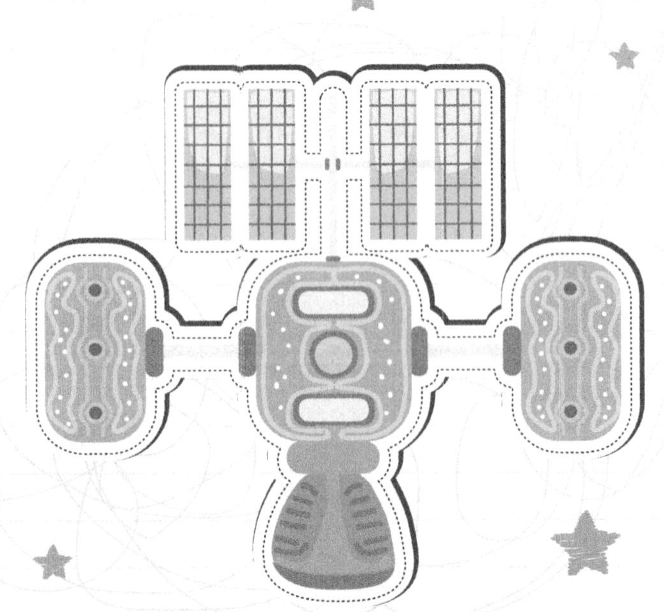

5. Fill in the sentence below with a verb that makes sense.
The baby .... all night long, keeping her parents awake.

6. Fill in the sentence below with a verb that makes sense.
She .... three marathons last year.

7. Write a sentence, using a past tense verb. Underline the verb. ....
....
....

8. Write a sentence, using a present tense verb. Underline the verb. ....
....
....

9. Write a sentence, using a future tense verb. Underline the verb. ....
....
....

10. Read the sentence below. Identify which verb tense is used.
Juan will play in a soccer game on Saturday.
Verb tense: ....
....

11. Using the sentence from #10, rewrite the sentence using the other two verb tenses.
Sentence 1: ....
....
Sentence 2: ....
....

12. Change the following sentence from present tense to past tense.
Raul is running down the street to catch up with his friends.
Past tense: ....
....

FITNESS PLANET

Let's get some fitness in! Go to page 167 to try some fitness activities.

FITNESS

Week 9 Science

Topic 1 Properties of Sound

In the following weeks, you are going to be exploring sound and how it travels. In order to better understand this concept, you need to be familiar with a few critical vocabulary words.

Pitch: how high or low a sound is.

Amplitude: how loud or soft a sound is.

Vibration: invisible waves that move up and down that our brains recognize as sounds.

Sound vibrations travel at different speeds, depending upon whether they are moving through a solid, liquid or gas (air).

Let's experiment with sound.

1. Use a pencil, wooden spoon or other object to tap a variety of solid objects (a table, door, countertop, bowl).
2. Experiment with the tapping to change the pitch and amplitude of the sound.
3. Now grab a second pencil, wooden spoon, etc. and tap them together in the air.
4. Again, experiment with changing the pitch and amplitude of the sound.

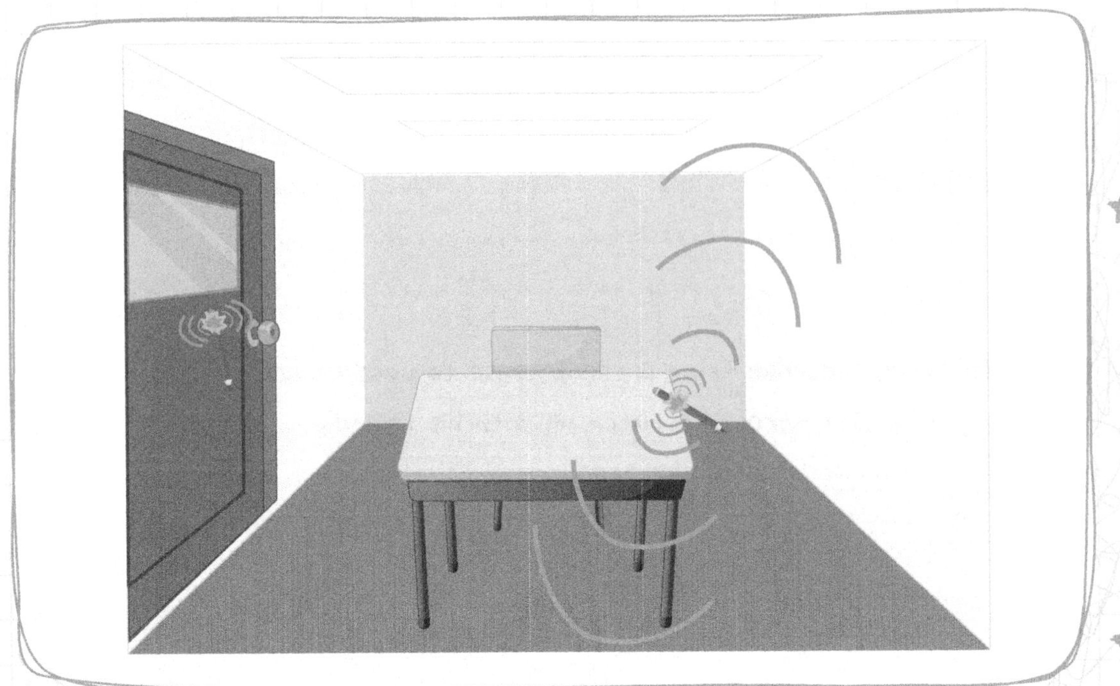

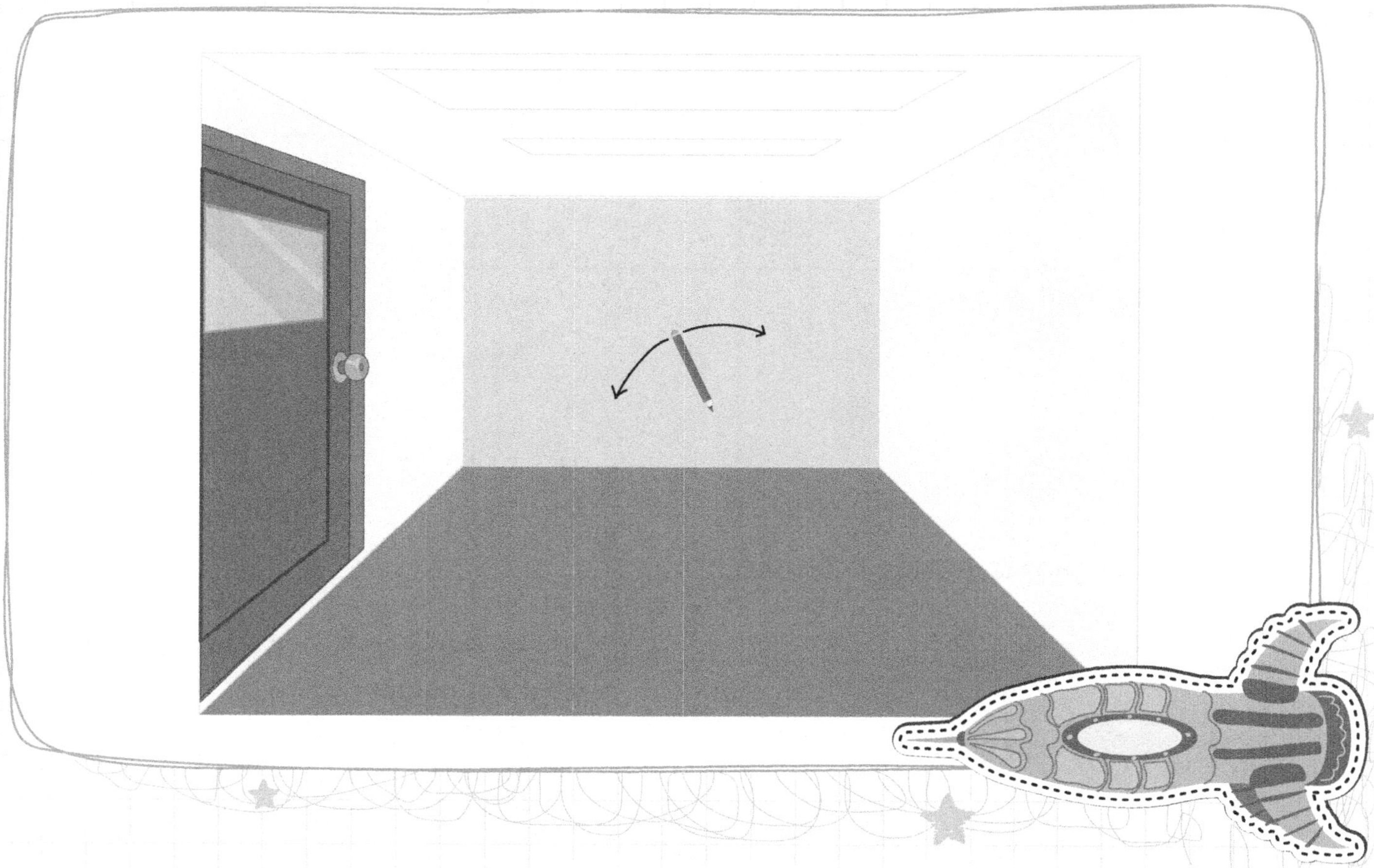

Now let's make a few predictions.

1. Do you think sound will travel fastest through solids, liquids or air? Why?

....

....

2. Why do you think sound is able to travel through all three states of matter?

....

....

3. What do you think you need to do to change the pitch or amplitude of a sound?

....

....

Week 9 Geometry

Topic 2 Dividing Rectangles

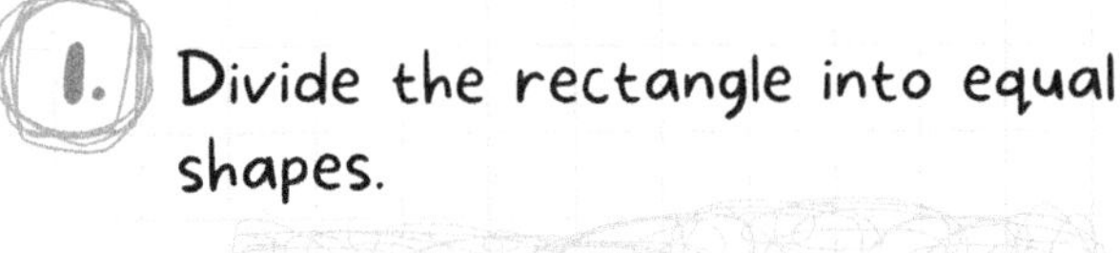

1. Divide the rectangle into equal shapes.

2. How many shapes is this rectangle divided into?

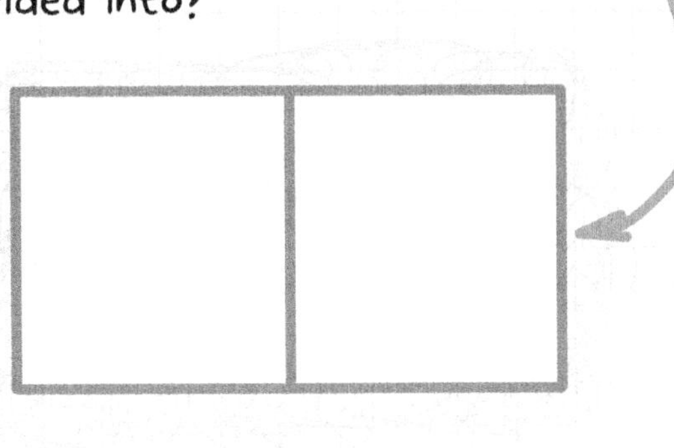

A. 4 B. 2 C. 6 D. 8

3. How many smaller squares make up this rectangle?

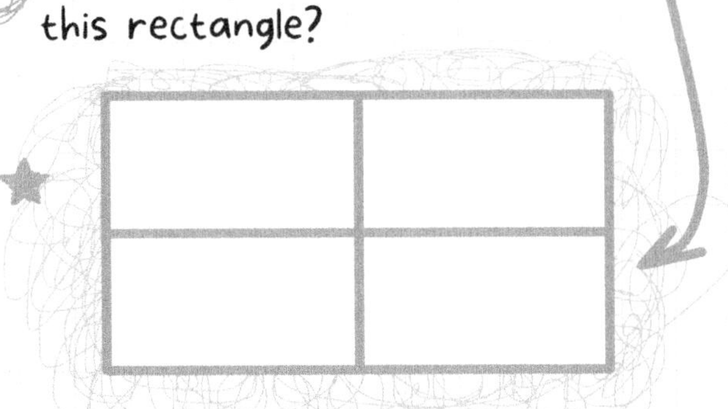

A. 4 B. 2 C. 6 D. 8

4. How many smaller squares make up this rectangle?

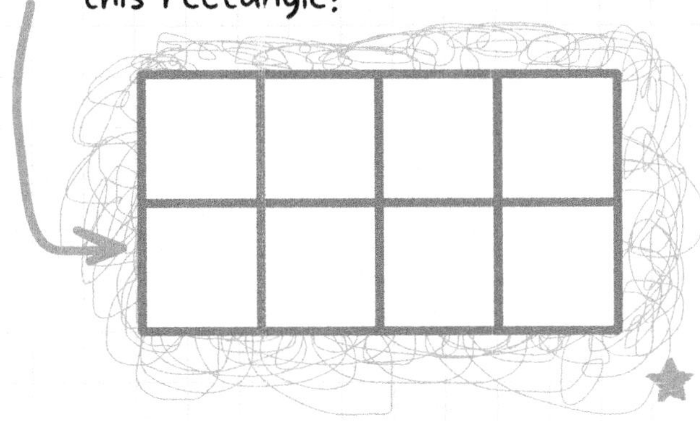

A. 4 B. 2 C. 6 D. 8

5. How many smaller squares make up this rectangle?

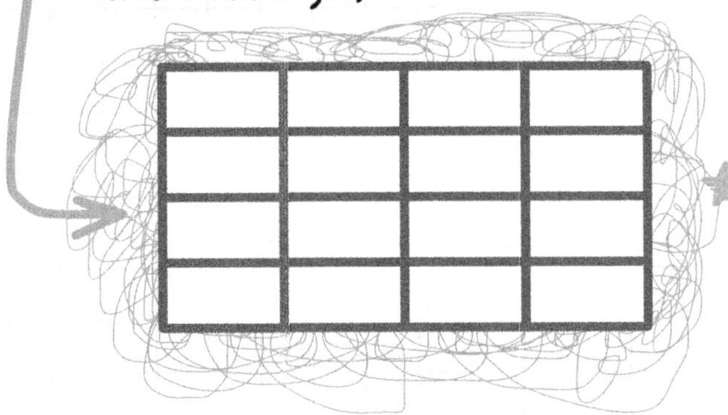

A. 14 B. 12 C. 16 D. 18

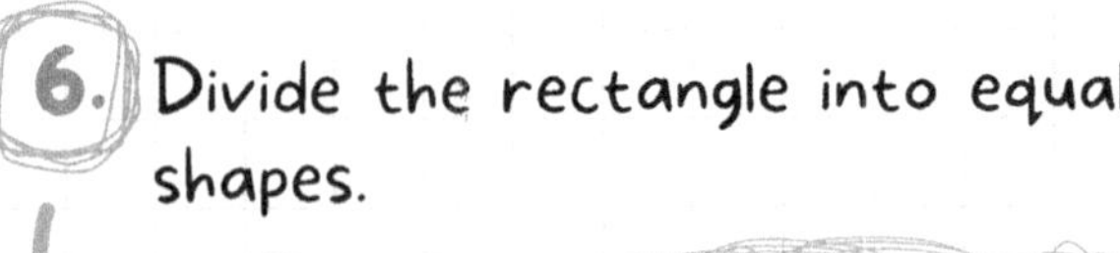

6. Divide the rectangle into equal shapes.

Week 9 Geometry

Topic 3 Dividing circles

1. Divide the circle into two equal parts.

2. How many shapes is this circle divided into?

A. 8 B. 10 C. 12 D. 14

3. Divide the circle into four equal parts?

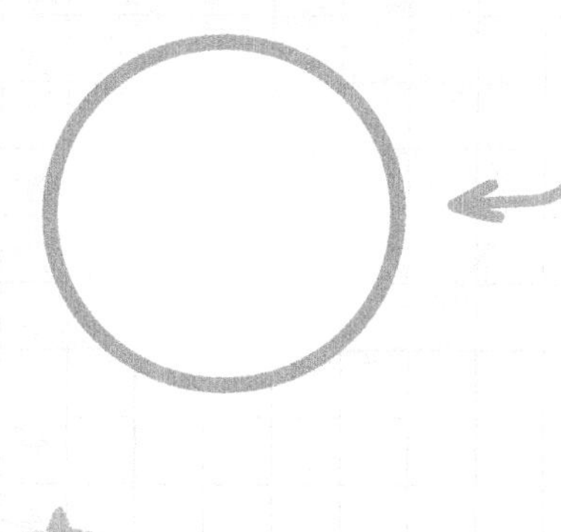

4. Divide the circle into six equal parts and shade in two parts.

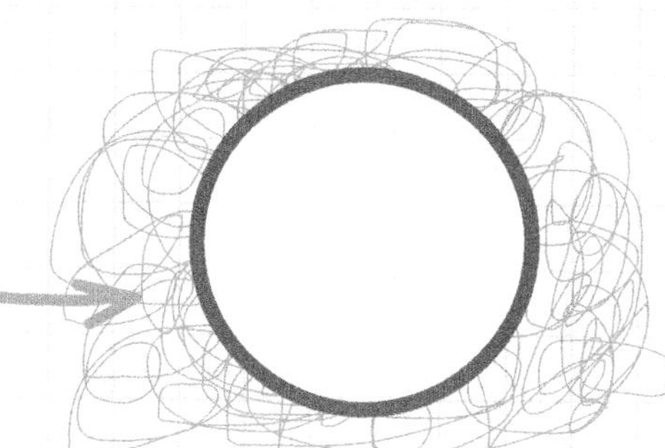

5. Divide the circle into four equal parts and shade in half.

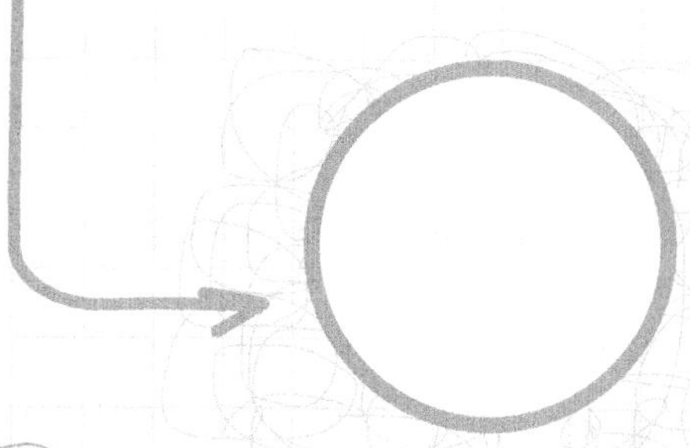

6. Divide the circle into three equal parts.

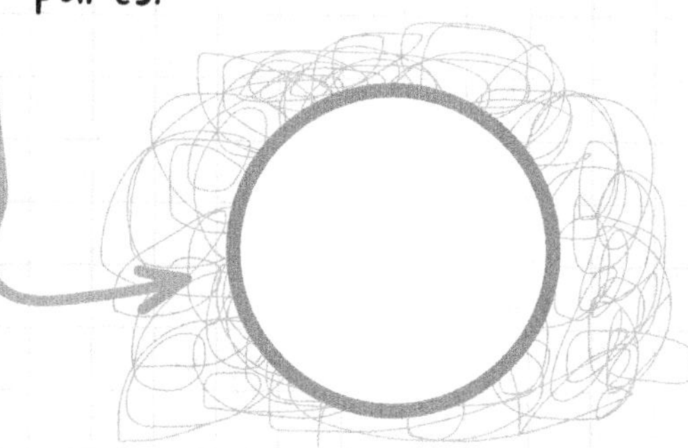

7. Divide the circle into eight equal parts and shade in half.

FITNESS PLANET ➡ Let's get some fitness in! Go to page 167 to try some fitness activities.

FITNESS

Today, you are going to review the elements of a story. Read the passage below and answer the questions that follow.

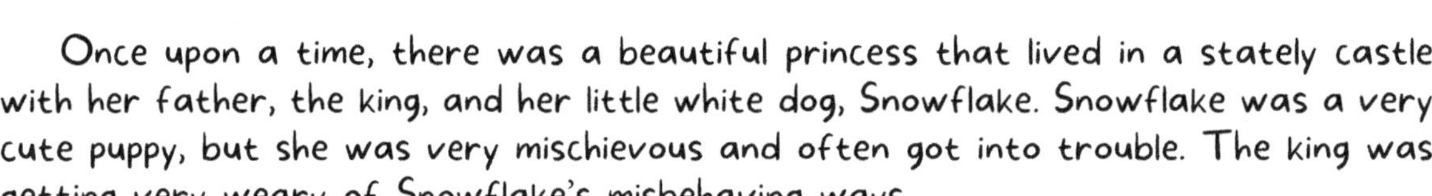

Once upon a time, there was a beautiful princess that lived in a stately castle with her father, the king, and her little white dog, Snowflake. Snowflake was a very cute puppy, but she was very mischievous and often got into trouble. The king was getting very weary of Snowflake's misbehaving ways.

One day, the princess was getting ready for a royal ball that was to begin at 7:00 that evening. A lot of very important people were going to be coming for the ball so it was important that the princess arrived on time. Right before the ball was set to begin, the princess went to get her tiara, but she couldn't find it! She searched all over her bedroom, but it was nowhere to be found. She just knew Snowflake had taken it!

She began searching all over the castle for Snowflake, but she couldn't find the naughty pup anywhere! Just as she was about to give up hope and go tell her father, she remembered seeing Snowflake sniffing mysteriously around the rose garden last week. She ran down the sweeping staircase and toward the door that led outside.

As she ran down the stone paths, she kept hoping to come across Snowflake. Finally, she turned the corner into the rose garden and saw Snowflake digging under one of the beautiful pink bushes.

"Snowflake, you naughty pup! Did you take my tiara?" the princess scolded.

Snowflake looked up at the princess and began barking.

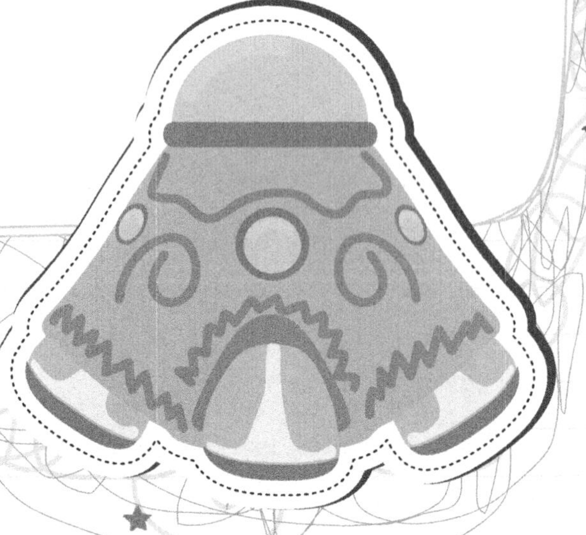

This story is incomplete. After reviewing basic story elements, you will finish the story.

1. List all the characters in the story.

2. Where does this story take place?
   - A. A house
   - B. A castle
   - C. At a party
   - D. In town

3. What event is taking place at 7:00 this evening in the story?
   - A. Dinner
   - B. A party
   - C. A ball

4. Describe the problem that is happening in the story.

5. Who does she suspect is to blame for the problem?

6. How does the princess know to look in the rose garden?

7. List two words that could be used to describe Snowflake.

8. On the lines below, finish the story. Be sure to include a solution to the problem, as well as an ending to the story.

Week 9 Geometry

Topic 4 Connecting shapes to fractions

1. How many smaller parts is this shape broken into?

A. 2 B. 3 C. 4 D. 5

2. How many smaller parts is this shape broken into?

A. 1 B. 2 C. 3 D. 4

3. How many smaller parts is this shape broken into?

A. 2 B. 3 C. 4 D. 5

4. How many smaller parts is this shape broken into?

A. 5 B. 4 C. 3 D. 2

5. How many smaller parts is this shape broken into?

A. 1 B. 2 C. 3 D. 4

6. Draw a rectangle and break it into four parts. Shade one part. What fraction represents the diagram?

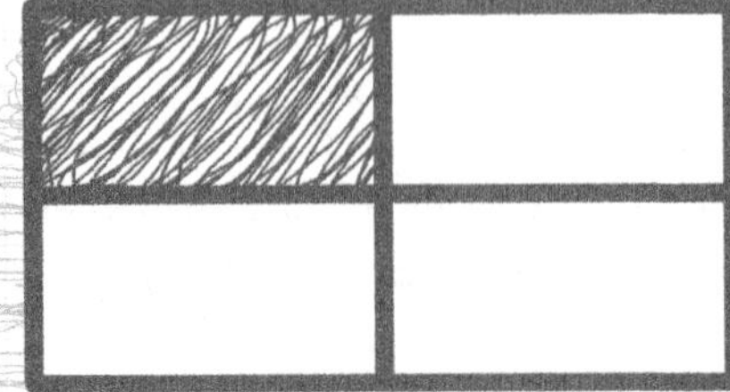

A. $\frac{1}{4}$ B. $\frac{1}{3}$ C. $\frac{1}{2}$ D. $\frac{1}{1}$

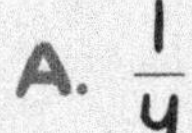

Today you will explore the different climate regions of the United States, as well as explain their characteristics.

There are 5 different climate regions in the United States. Research each one listed below and write a brief summary of the climate characteristics of each. Be sure to use complete sentences.

1. **Northeast**

2. **Southeast**

3. **West**

4. **Southwest**

5. **Midwest**

FITNESS PLANET → Let's get some fitness in! Go to page 167 to try some fitness activities.

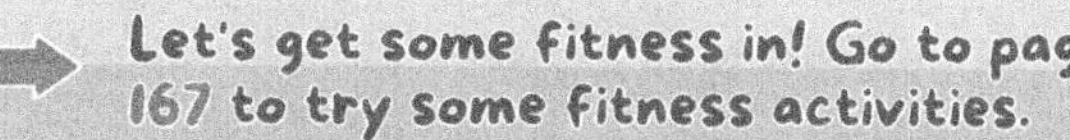

Grade 2-3

WEEK 10

Let's see what you know about:

- quotation marks
- sound travel
- informational texts
- economics and more!

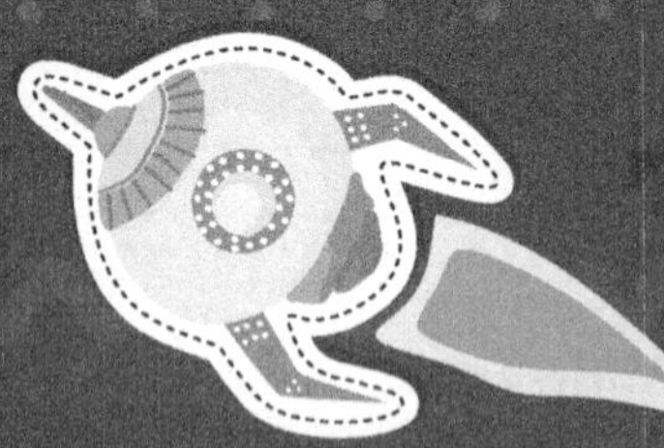

Week 10 More Calculations

Topic 1 Addition and subtraction review

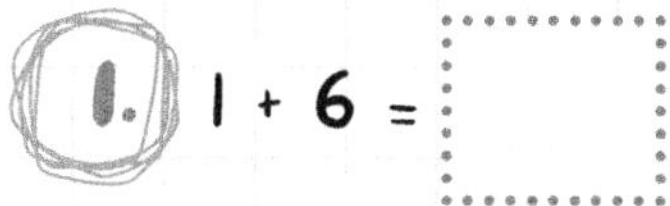

1. 1 + 6 = ☐

A. 6 B. 7 C. 8 D. 9

2. 3 + 2 = ☐

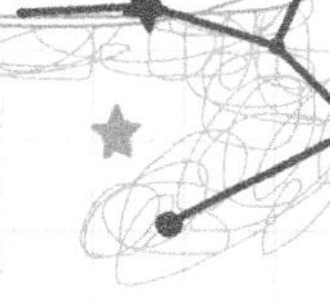

A. 2 B. 3 C. 4 D. 5

3. 5 + 8 = ☐

A. 11 B. 12 C. 13 D. 14

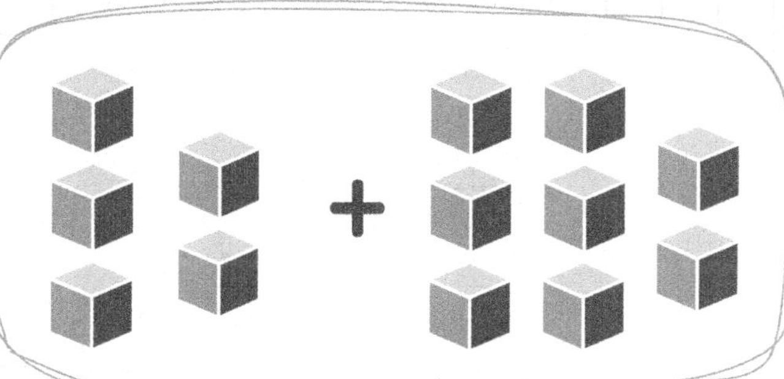

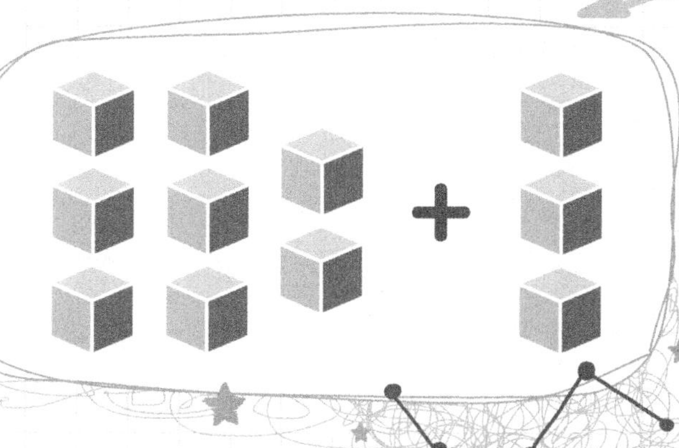

4. 8 + 3 = ☐

A. 9 B. 10 C. 11 D. 12

5. 2 + 7 = ☐

A. 8 B. 9 C. 10 D. 11

6. 6 + 5 = ☐

A. 8 B. 9 C. 10 D. 11

Week 10 More Calculations

Topic 1 Addition and subtraction review

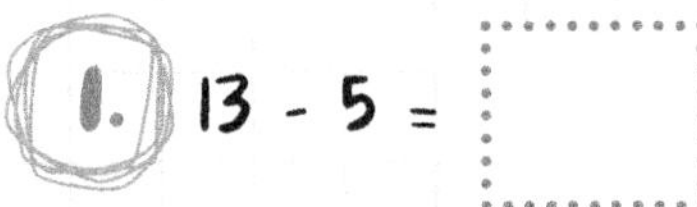

1. 13 - 5 = ☐

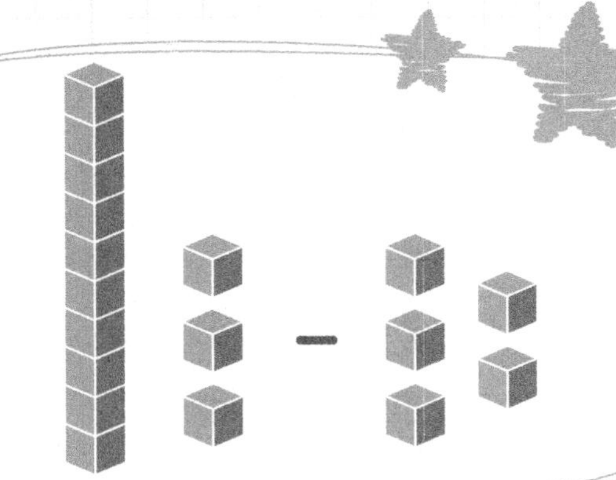

A. 6 B. 7 C. 8 D. 9

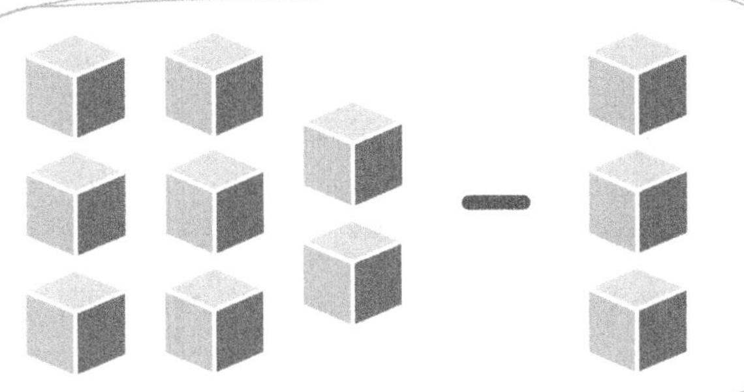

2. 8 - 3 = ☐

A. 4 B. 5 C. 6 D. 7

3. 15 - 8 = ☐

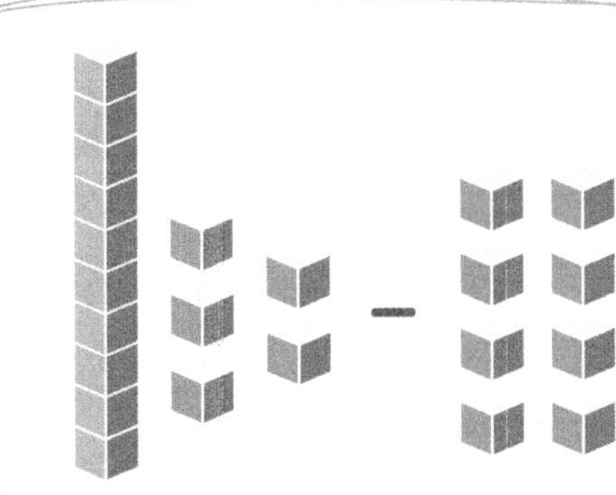

A. 4 B. 5 C. 6 D. 7

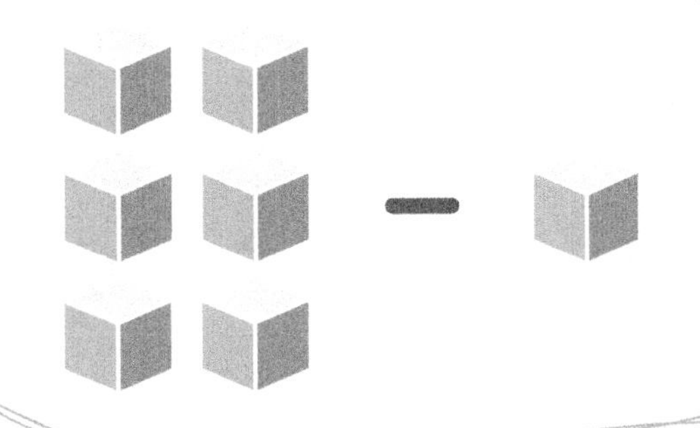

4. 6 - 1 = ☐

A. 4 B. 5 C. 6 D. 7

5. 11 - 6 = ☐

A. 3 B. 4 C. 5 D. 6

6. 16 - 7 = ☐

A. 9 B. 8 C. 7 D. 6

Week 10 Language Review

Topic 1 Quotation Marks

Quotation marks are used to indicate when a person is speaking in text. To correctly use quotations in your writing, it is important to follow these rules:

1. **The quotation marks only go around the words that are being spoken.**

   Example: "The weather is beautiful today," said Lora.

   Notice that the words *said Lora* are not in quotes. This is because she does not actually say these words aloud. They are meant to tell the reader who is speaking.

2. **Use commas to separate the spoken words from the text that tells the reader who is speaking.**

   Example: "I am so excited to go outside and play," Lora exclaimed.

   Notice that a comma is also used to separate the spoken words from the text that tells the reader who is speaking.

3. **Punctuation marks such as commas, exclamation marks and question marks go inside the quotation marks.**

   Example: "What would you like to do today?" her mom asked.
   Notice that the question mark in this example, and the comma in the example above, are placed inside the quotation marks. Punctuation marks should not be placed on the outside of the quotation marks. Question marks and exclamation marks are used in place of the comma, as seen in the example above.

4. **One set of quotation marks can be used when a person is speaking several sentences at a time.**

   Example: "I just love fall. The leaves turn splendid colors and the air is crisp. Everybody bakes pumpkin and apple treats. Plus, Halloween is my very favorite holiday," stated Mrs. Johnson.

   Notice in this example, the speaker states four sentences and only one set of quotation marks is used. You do not need to place quotation marks around every individual sentence, if they are all spoken together by one person.

5. **A sentence can be separated into two parts with quotation marks around each part that is spoken aloud and a comma before and after the part that tells the reader who is speaking.**

   Example: "This," exclaimed Luis, "is my favorite day ever!"

   Notice that each portion of the sentence that is spoken aloud is surrounded by quotation marks.

Week 10 Language Review

Topic 1 Quotation Marks

Now it's your turn to practice using quotation marks in writing. Write the quotation marks and other punctuation marks in each sentence where they belong.

1. The weather is turning colder Montrell said.

2. Yikes! I'm scared of spiders she shouted.

3. Is it lunchtime yet he asked the teacher.

4. I am so excited about this weekend. My mom said we could go to the store to buy a new dress for the party. I hope I find a pink one Joelle told Michelle.

5. No screamed Marco.

6. Did you bring a coat the teacher asked.

7. This field trip Hugo exclaimed is going to be the best one yet!

8. What should I do next she asked her dad.

9. Let's go mom shouted up the stairs.

10. On Thursday mom stated we will go to the store.

Today, you are going to test how sound travels. When sounds are made, waves are created that travel through the air to our ears. These are called **sound waves.**

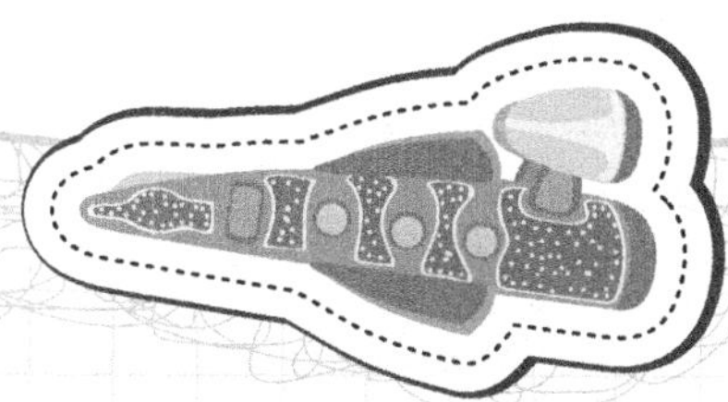

Materials Needed:

* Ruler
* 2 different sized spoons
* Approximately 4 feet of yarn

Procedure:

1. Tie a loop in the middle of the yarn and insert the handle of one of the spoons. Pull the yarn tightly so that the spoon hangs. You should have approximately equal lengths of yarn on each side of the spoon.
2. Pick up the yarn and wrap each side around the pointer finger on each hand.
3. Hold each side of the yarn to your ears...not in your ears, just next to them. The spoon should hang below your waist.
4. Have someone gently hit the spoon, using the ruler.
5. Experiment with different sized spoons and hitting the spoon harder and softer with the ruler.

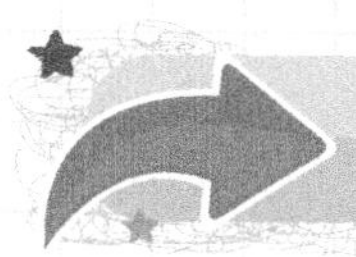

FITNESS PLANET

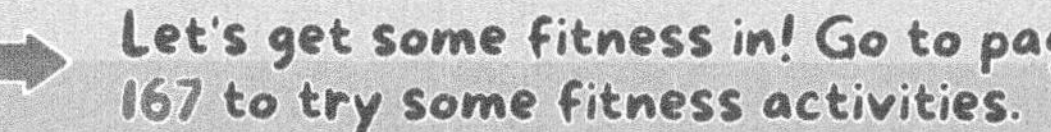

Let's get some fitness in! Go to page 167 to try some fitness activities.

Follow-Up Questions:

1. What happened when the ruler hit the spoon?

2. How do you think this happens?

3. What happens when you use different sized spoons?

4. Can anyone else in the room hear a sound when the ruler hits the spoon? Why or why not?

Topic 2 Addition and subtraction review

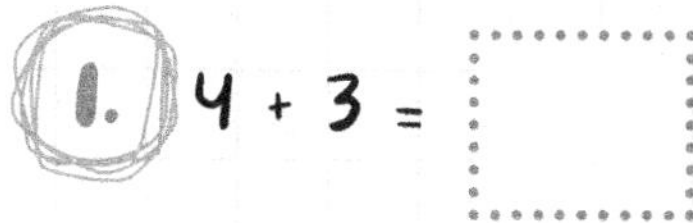

1. 4 + 3 = ☐

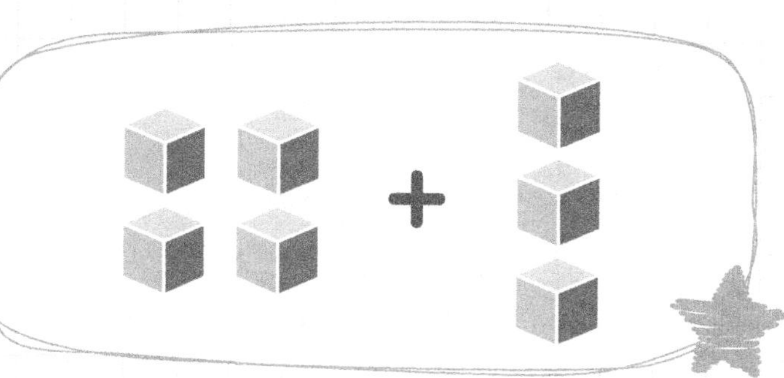

A. 6 B. 7 C. 8 D. 9

2. 7 - 5 = ☐

A. 2 B. 3 C. 4 D. 5

3. 2 + 8 = ☐

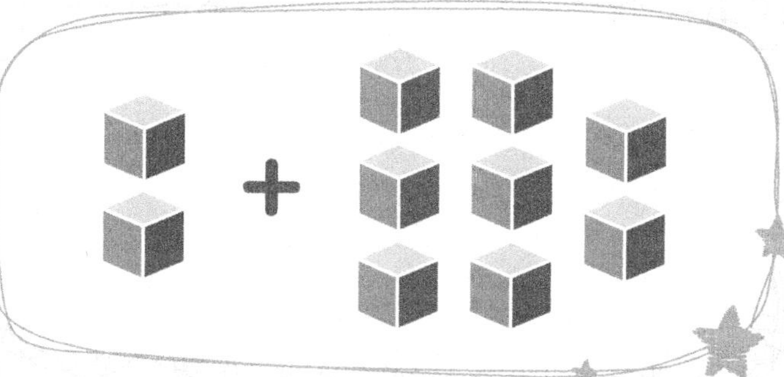

A. 7 B. 8 C. 9 D. 10

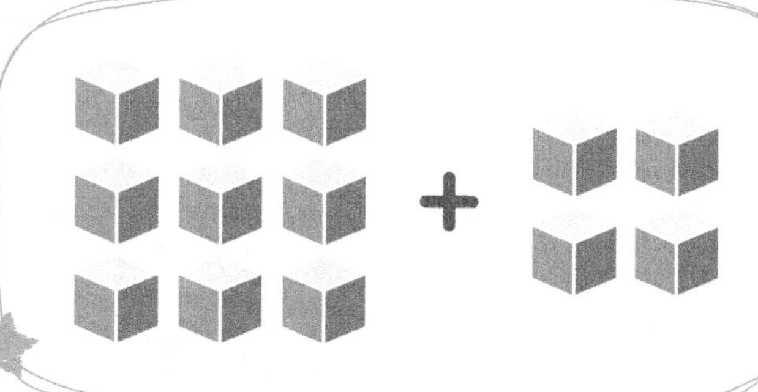

4. 9 + 4 = ☐

A. 11 B. 12 C. 13 D. 14

5. 6 - 2 = ☐

A. 4 B. 5 C. 6 D. 7

6. 1 - 1 = ☐

A. 3 B. 2 C. 1 D. 0

Topic 2 Addition and subtraction review

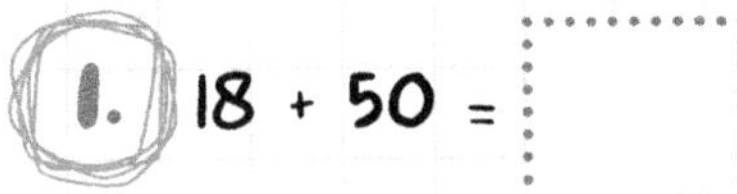

1. 18 + 50 = ☐

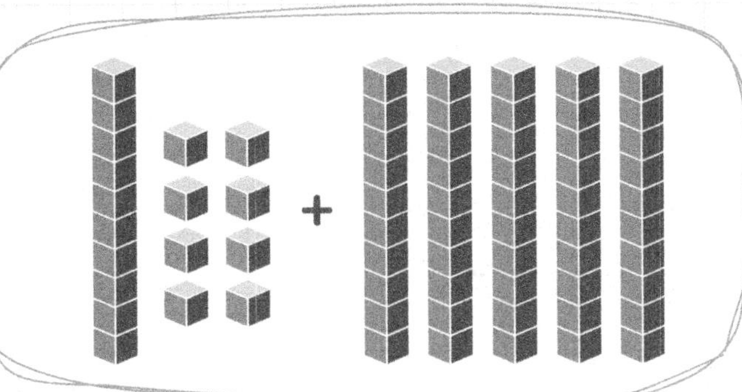

A. 64 B. 66 C. 68 D. 70

2. 20 + 39 = ☐

A. 57 B. 59 C. 61 D. 63

3. 41 + 11 = ☐

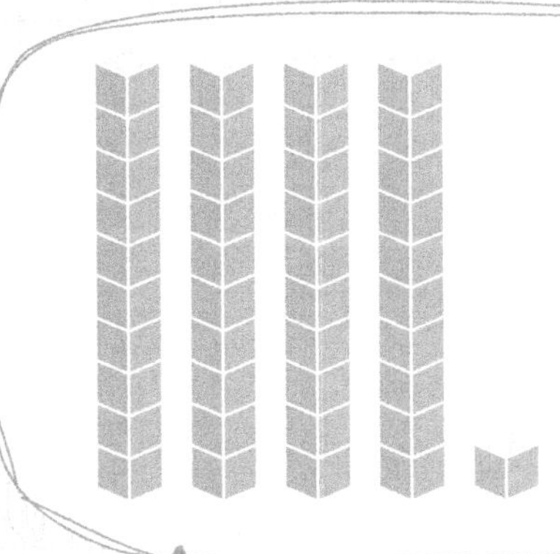

A. 41 B. 51 C. 52 D. 62

4. 15 + 74 = ☐

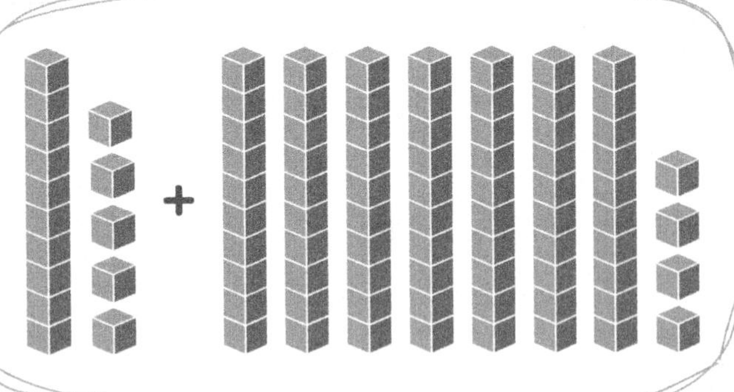

A. 89 B. 90 C. 91 D. 92

5. 63 + 35 = ☐

A. 96 B. 97 C. 98 D. 99

6. 17 + 87 = ☐

A. 104 B. 94 C. 84 D. 74

Today, you are going to compare and contrast the main idea and details of two texts on the same topic. Read the passages below.

Passage 1

Plants are living things that are very important to our ecosystem. Plants provide many benefits to humans and animals. They give us oxygen and help improve our air quality. Bees need flowering plants for pollination. Plants also provide food and shelter to many insects and animals. Without plants, humans and animals would not have the things they need to survive.

Passage 2

Growing plants is not too tricky! First, you need seeds, water, an area that gets lots of sunlight and good soil. You will need to dig a small hole in the soil and place the seed gently inside. Cover it back up with soil so that it does not blow away or get eaten by an animal. Be sure to water the seed everyday. As long as you provide sunlight, water, air and nutrients, your seed should have no problem growing!

Now you will compare and contrast the two passages by answering the questions below.

1. Identify the main idea of each passage below.

Passage 1: ....

....

....

Passage 2: ....

....

....

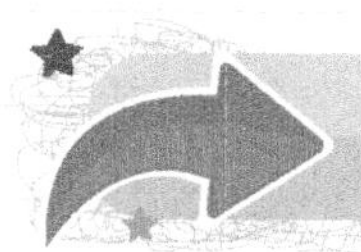

FITNESS PLANET

Let's get some fitness in! Go to page 167 to try some fitness activities.

2. Identify two key details from each passage below.

Passage 1: ....

Passage 2: ....

3. How are these two passages similar?

4. How are these two passages different?

5. What can you conclude about plants after reading these two passages?

6. How might animals use plants for shelter?

7. How can we protect plants and animals in our ecosystem?

Week 10 More Calculations

Topic 3 Addition and subtraction review

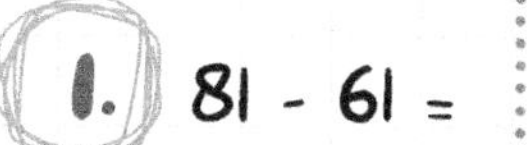

1. 81 - 61 = ☐ ....

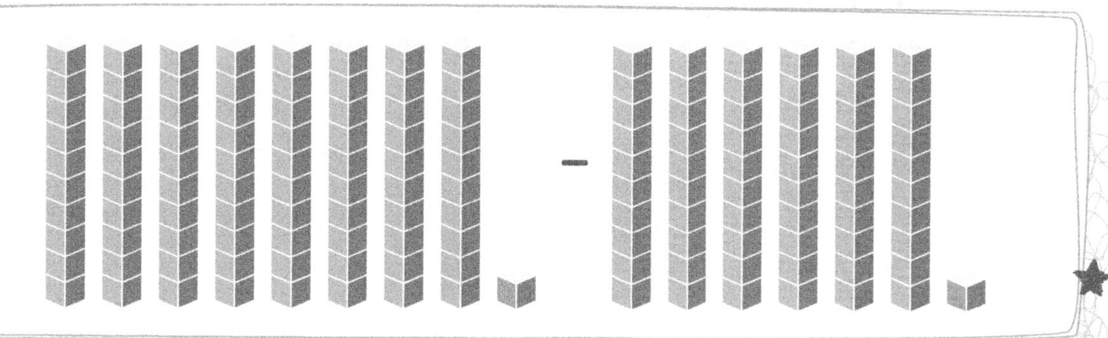

A. 10 B. 20 C. 30 D. 40

2. 67 - 32 = ☐ ....

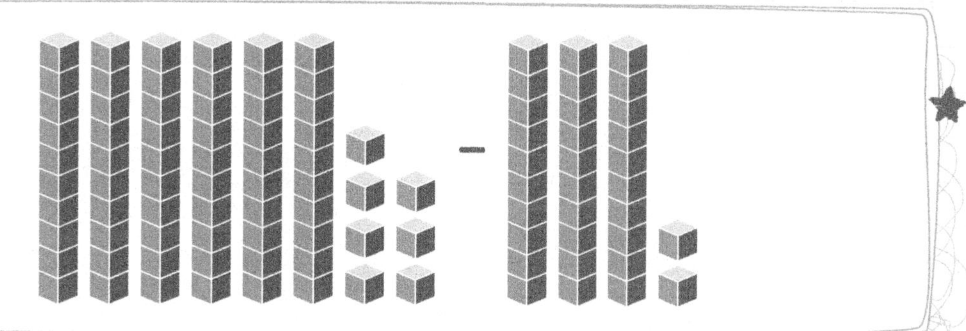

A. 25 B. 35 C. 45 D. 55

3. 95 - 55 = ☐ ....

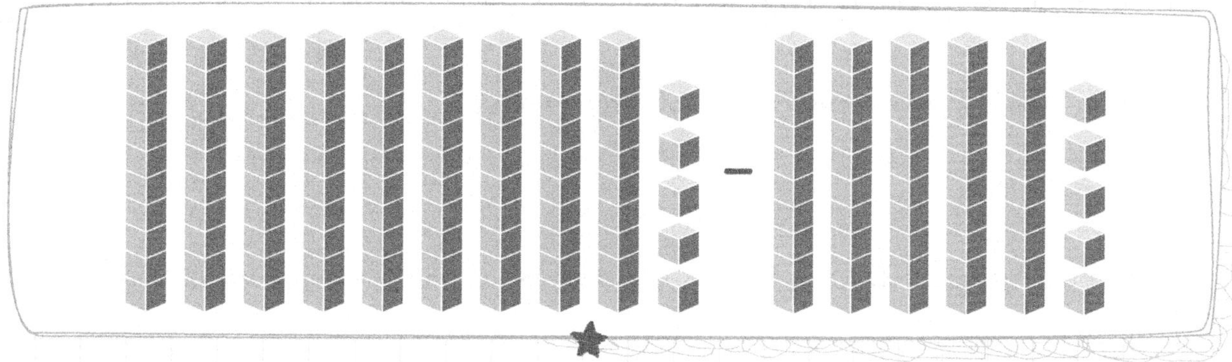

A. 20 B. 30 C. 40 D. 50

Topic 3 Exploring Economics in Your Community

Today, you will review the concept of goods and services by exploring the specific ones available in your own community.

Remember, goods are items that are tangible while services are intangible.

Think about the local businesses and industries that are located in your community to complete the chart below. Be specific in your examples.

Goods	Services

Now answer the following questions about the goods and services in your community.

1. What are the main industries in your community?

....

....

2. How have these industries had an economic impact on the community?

....

....

Get ready to master:

* addition & subtraction
* context clues
* number patterns and more!

Week 11 Making Calculations Mixed Review

Topic 1 Odd and even numbers

1. 15 ☐ Even ☐ Odd

2. 1 ☐ Even ☐ Odd

3. 18 ☐ Even ☐ Odd

4. 20 ☐ Even ☐ Odd

5. 4 ☐ Even ☐ Odd

6. 12 ☐ Even ☐ Odd

7. 6 ☐ Even ☐ Odd

8. 58 ☐ Even ☐ Odd

9. 31 ☐ Even ☐ Odd

10. 2 ☐ Even ☐ Odd

Week 11 Making Calculations Mixed Review

Topic 1 Addition and subtraction review

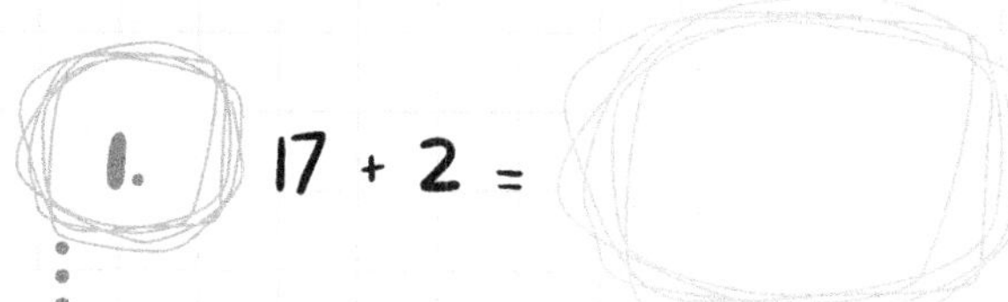

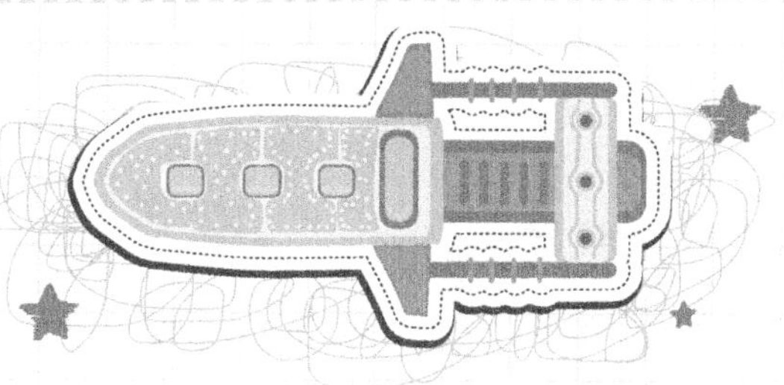

1. 17 + 2 =

2. 5 + 11 =

3. 18 + 1 =

4. 2 + 14 =

5. 15 + 2 =

6. 20 - 3 =

7. 14 - 4 =

8. 14 - 9 =

9. 10 - 6 =

10. 17 - 5 =

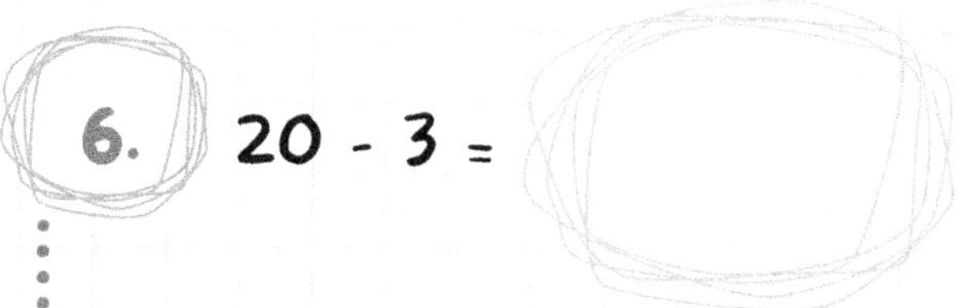

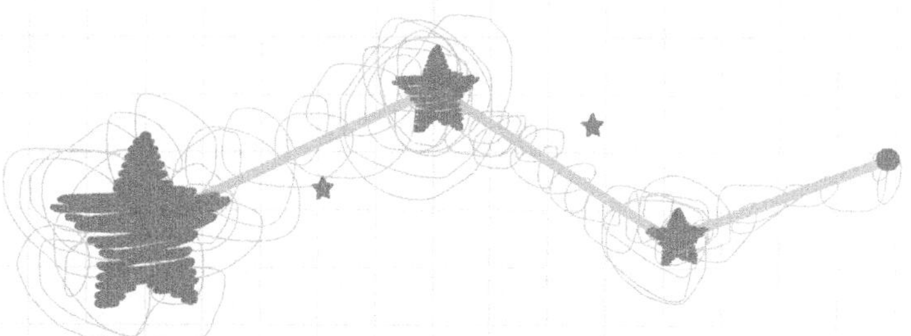

FITNESS PLANET

Let's get some fitness in! Go to page 167 to try some fitness activities.

Week 11 Language Review

Topic 1 Context Clues

Context clues are hints that an author gives to a reader to help define a difficult or unfamiliar word.

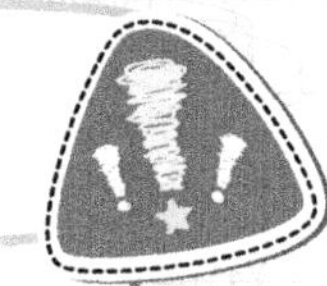

There are several types of context clues that can be used in texts. Take a look at the five main types of context clues, as well as examples, outlined below. In the example sentences, the unfamiliar word is underlined while the context clue is highlighted.

1. **Definition** - the unknown word is defined in the sentence or surrounding sentences.

   Example: The Civil War was a significant, or important, part of our country's history.

2. **Example** - specific examples are given in the sentence as clues to the meaning of the unknown word.

   Example: Hors d'oeuvres, including fruit, shrimp and cheese and crackers, were served at the party.

3. **Synonym** - another word with the same meaning as the unknown word is also used in the sentence.

   Example: The bright light from the sun during the eclipse was blinding.

4. **Antonym** - a word with the opposite meaning of the unknown word is used in the sentence.

   Example: The vast city was so different than the small town he was used to living in.

5. **Inference** - the meaning is not given so you use context clues to make a guess about the meaning of the unknown word.

   Example: The new student was so arrogant. He bragged about himself all day long.

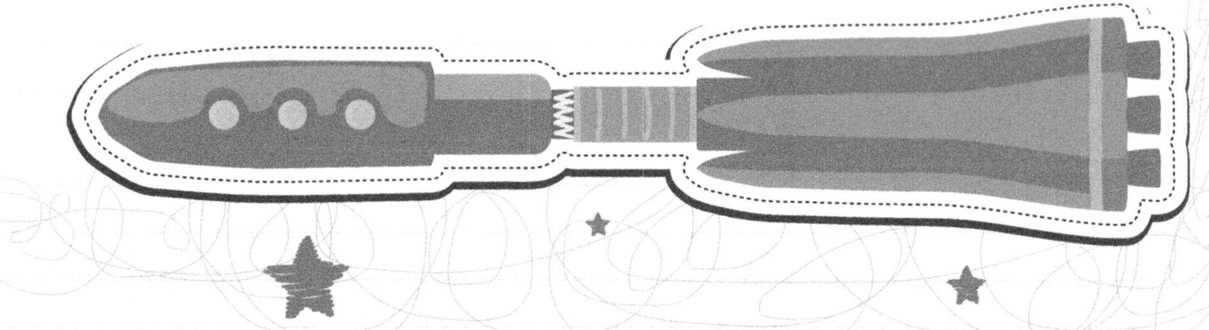

Week 11 Language Review

Topic 1 Context Clues

Now it's your turn to practice using context clues to determine the meaning of an unknown word. Write your guess below each sentence.

1. She was hoping the coin was <u>valuable</u>, but her dad told her it was worthless.

2. His new teacher, Ms. Smith, was quite <u>affable,</u> or kind.

3. The president of the company needed a break so he took a <u>hiatus</u> from his job.

Read each sentence below and determine which type of context clue is used in the sentence.

1. When their mom caught them sneaking cookies after dinner, she <u>admonished</u> them and they knew they were in trouble.
   A. Inference
   B. Definition
   C. Example
   D. Synonym
   E. Antonym

2. The day of the funeral was a very <u>somber,</u> or sad, day.
   A. Inference
   B. Definition
   C. Example
   D. Synonym
   E. Antonym

3. They went to the zoo on a particularly <u>sultry</u> day. They were so hot and sweaty and drank several bottles of water each.
   A. Inference
   B. Definition
   C. Example
   D. Synonym
   E. Antonym

Week 11 Science

Topic 1 Can Sound Travel Through Liquids?

Today, you are going to investigate whether or not sound can travel through a liquid, such as water, and compare it to how sound travels through air.

Materials Needed:

* 2 metal spoons
* Plastic 2 liter bottle, no lid
* Scissors
* Large bowl of water

Procedure:

1. Place the bowl of water on a table.
2. Carefully cut the bottom quarter of the 2 liter bottle off.
3. Place the cut side of the bottle into the bowl of water and place your ear against the hole on the other side.
4. Listen carefully, as you have another person tap the two metal spoons together under the water.

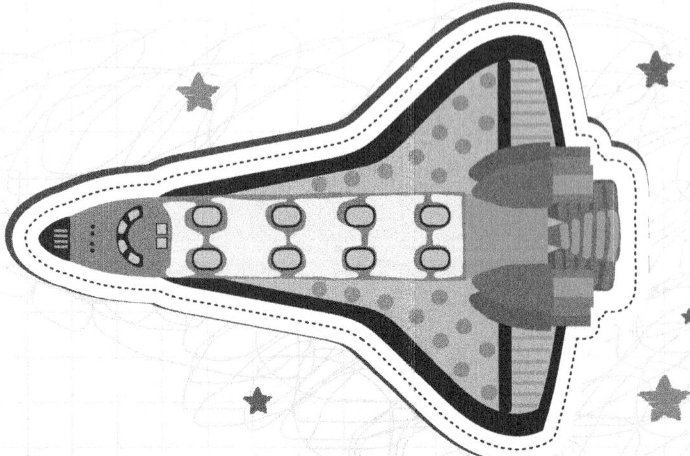

Follow-Up Questions:

1. What did you hear as the spoons were being tapped together?

2. Why do you think this happened?

3. Do you think sound travels faster through air or water? Why?

FITNESS PLANET

Let's get some fitness in! Go to page 167 to try some fitness activities.

Topic 2 Two-digit addition and subtraction review

1. 43 + 68 = ☐

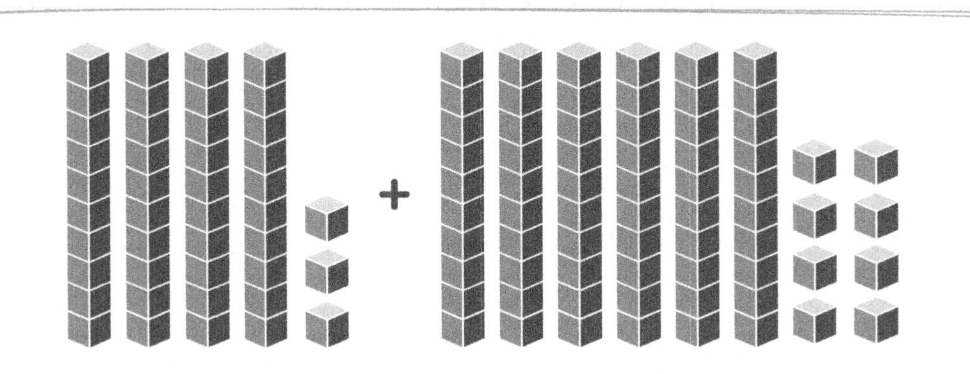

A. 109　B. 111　C. 113　D. 115

2. 26 + 92 = ☐

A. 114　B. 116　C. 118　D. 120

3. 67 + 10 = ☐

A. 77　B. 87　C. 97　D. 107

4. 91 + 34 = ☐

A. 123　B. 125　C. 127　D. 129

5. 25 + 23 = ☐

A. 30　B. 38　C. 48　D. 50

6. 59 - 47 = ☐

A. 10　B. 12　C. 14　D. 16

7. 89 - 64 = ☐

A. 23　B. 25　C. 27　D. 29

What is the next number of the pattern:

1. 5 → 10 → 15 → ______
2. 200 → 300 → 400 → ______
3. 130 → 140 → 150 → ______
4. 700 → 710 → 720 → ______
5. 60 → 65 → 75 → ______
6. 40 → 50 → 60 → ______
7. 375 → 385 → 395 → ______
8. 280 → 290 → 300 → ______
9. 80 → 70 → 60 → ______
10. 990 → 985 → 980 → ______

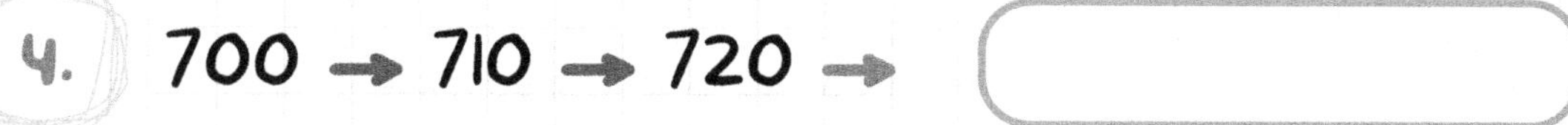

Week 11 Reading Passage

Topic 2 Literature

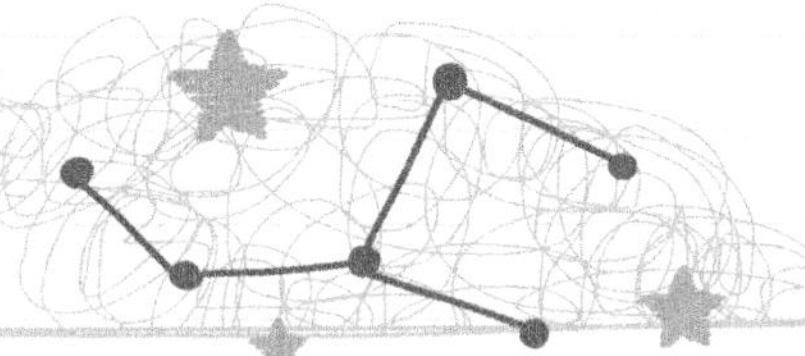

Read the passage below.

Micah loves playing soccer. He started playing when he was just four years old and has played every year since. This year, he is on Team Jamaica. Most of the other players on the team are new to the sport. He knows that he is the best player on the team, but he doesn't like to brag. Instead, he tries to help the other kids improve.

He also really likes his coach, Coach Damien. Coach Damien knows a lot about soccer and teaches Micah many new drills that he has never done before. Micah enjoys going to practice each week. Games are his favorite, though. Micah loves the feeling he gets when he scores a goal during a game and the whole crowd cheers.

Micah works hard during practices because he really wants to play soccer in high school and college. He even dreams of becoming a professional soccer player, but he knows he has to be diligent about practicing and improving. He will work hard, though, to reach his goals.

Answer the following questions about the passage.

1. What do you think the word brag means in the passage?

   A. To be the best
   B. To tell others how good you are
   C. To make fun of others

2. What do you think the word diligent means in the passage?

   A. Hard-working
   B. Forgetful
   C. Hopeless

3. Why does Micah try to teach the other players in the passage?

....

....

....

....

In the passage you just read, Micah was passionate about soccer and dreamed of becoming a professional soccer player one day. On the lines below, write a short informative piece about what you dream of becoming when you are older. Be sure to include a topic sentence, details to support your topic and a concluding statement. Write in complete sentences, using appropriate capitalization, punctuation and spelling.

FITNESS PLANET
Let's get some fitness in! Go to page 167 to try some fitness activities.

Week 11 Making Calculations Mixed Review

Topic 3 Addition and subtraction under 1000

1. 750 + 10 = ...

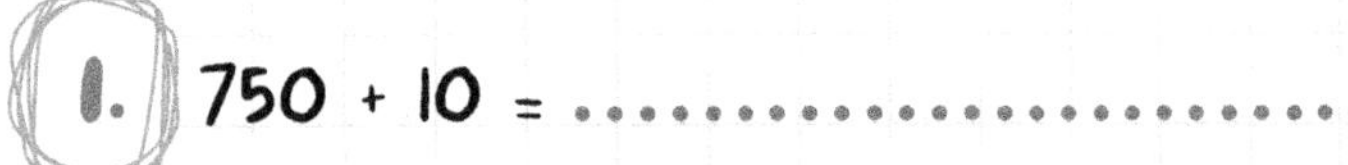

A. 750 B. 760 C. 770 D. 780

2. 520 - 40= ....

A. 440 B. 460 C. 480 D. 500

3. 80 + 30 = ...

A. 100 B. 110 C. 120 D. 130

4. 220 + 60 = ...

A. 260 B. 280 C. 300 D. 320

5. 90 - 20= ....

A. 50 B. 60 C. 70 D. 80

6. Our class has 12 students who ride the bus and 4 students who walk to school. How many more students ride the bus?

....

A. 8 B. 4 C. 16 D. 12

7. We planted flowers in our garden. We planted 5 blue flowers and 12 red flowers. How many flowers will be in our garden?

....

A. 17 B. 16 C. 15 D. 14

Topic 3 Exploring Economics in Your Community

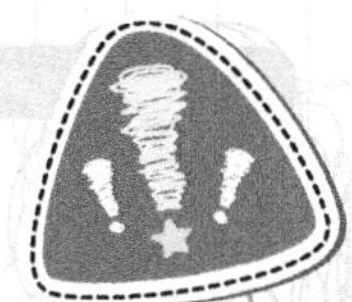

Consumers are people who purchase goods and services for their own use.

Producers are people or companies who make or supply goods to consumers.

Before making a purchasing decision, consumers must weigh the costs and benefits of their purchase.

Read the scenario below. Then, complete the chart that follows.

Olivia wants her family to buy a dog. Before agreeing to do so, her mom says she must make a list of the costs and benefits of owning a dog. Help Olivia by reading the sample and then completing the chart below.

Costs	Benefits
Veterinary bills	A lovable companion

1. Do you think Olivia's family should purchase a dog? Why or why not?

....

....

....

....

Grade 2-3

WEEK 12

Let's make a big finish with:

- measurement and time
- shades of meaning
- and how sound travels through objects!

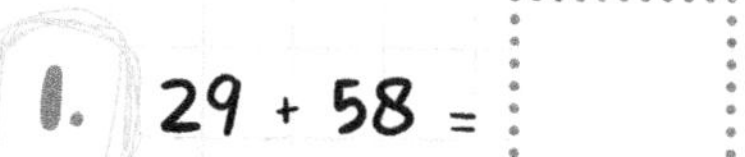

1. 29 + 58 = ☐ ....

A. 67 B. 77 C. 87 D. 97

2. What unit would you use to measure a clipboard?

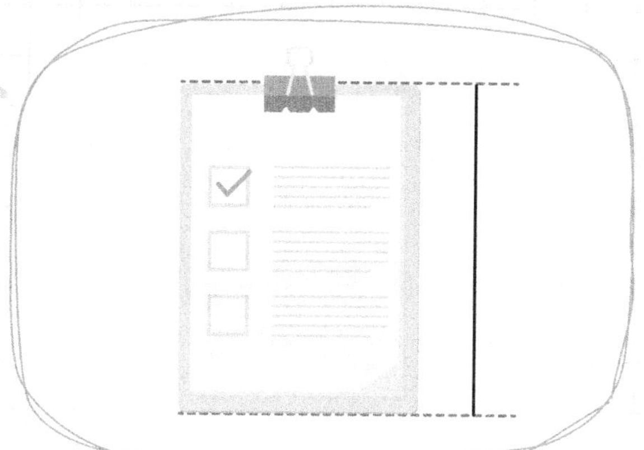

....

A. inch B. foot C. mile D. yard

3. Choose an item around you. ....

What is an estimate of its length? ....

4. Which number is not odd?

....

A. 1 B. 9 C. 6 D. 7

5. Which number is not even?

....

A. 14 B. 27 C. 28 D. 52

6. 12 - 7 =

....

A. 5 B. 6 C. 7 D. 8

7. Show 1:38 on the clock.

....

....

....

FITNESS PLANET

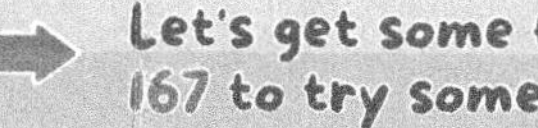

Let's get some fitness in! Go to page 167 to try some fitness activities.

FITNESS

Week 12 Mixed Review

Topic 1 Numbers, operations, measurement and time

1. What would be an appropriate time to go for a walk?

A. 2:00 AM B. 2:00 PM

2. 32 + 88 =

$$\begin{array}{r} 32 \\ +\ 88 \\ \hline \end{array}$$

A. 110 B. 120 C. 130 D. 140

3. What would you use to measure a banana?

A. Kilometer C. centimeter
B. meter D. milimeter

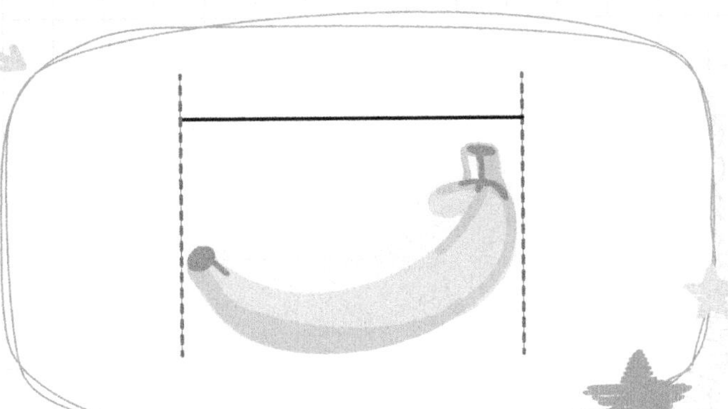

4. 15 - 6 =

A. 8 B. 9 C. 10 D. 22

5. 14 + 11 =

$$\begin{array}{r} 14 \\ +\ 11 \\ \hline \end{array}$$

A. 24 B. 15 C. 25 D. 30

6. Jonah has a collection of stuffed animals. He has 16 animals and he receives 8 more for his birthday. How many stuffed animals does he have after his birthday?

A. 24 B. 26 C. 28 D. 30

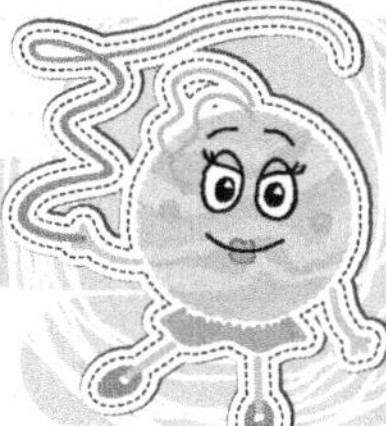

Shades of meaning refers to the slight differences in meaning between similar words or phrases.

Example: Insert each of the words from the table into the sentence below to see how the meaning of the sentence changes slightly with each different word.

The jury .... the man was innocent of the crime.

knew
believed
suspected
heard
wondered

Now you will practice brainstorming words that are slightly different in meaning than the given word. Write your words in the tables below.

happy	said	walked

Choose one set of words from the table above and create one sentence that each word can be inserted into (as in the example above). Write each sentence on the lines below.

1. ....

2. ....

3. ....

4. ....

5. ....

6. ....

Explain how the meaning of the sentence changes with each different word used.

....

FITNESS PLANET → Let's get some fitness in! Go to page 167 to try some fitness activities.

FITNESS

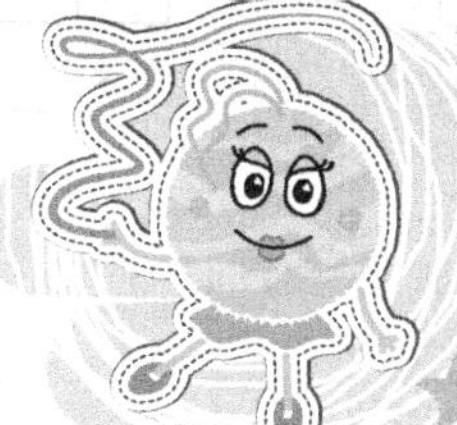

Week 12 Science

Topic 1 How Does Sound Travel Through Solids?

Today, you are going to explore how sound travels through solids, compared to air and liquids, as you investigated in previous weeks.

Materials Needed:

* 2 small plastic yogurt containers
* Scissors
* String
* Tape
* A partner

Procedure:

1. Using scissors, poke a hole in the bottom of each yogurt container.
2. Find a place where you have plenty of open space (for example, on the sidewalk or in the yard).
3. Talk to your partner in a normal speaking voice. Back up until you can no longer hear one another when talking.
4. Hold your container up to your mouth and speak into it. Have your partner do the same.
5. Now, cut a length of string long enough to reach from you to your partner.
6. Poke each end of the string through the hole on the bottom of the containers and secure with tape.
7. Hold the containers firmly and gently back up until the string is stretched tightly.
8. Take turns speaking to one another while the other person holds the container up to their ear.

Topic 1 How Does Sound Travel Through Solids?

Follow-Up Questions:

1. Were you able to hear your partner when you first began speaking into the container? Why or why not?

2. Were you able to hear your partner when you added the string and spoke into the container? Why or why not?

3. Do you think sound travels faster through air, water or solids? Why or why not?

Week 12 Mixed Review

Topic 2 Time, measurement and money

1. Which item is shorter, a lamp or a Christmas tree?

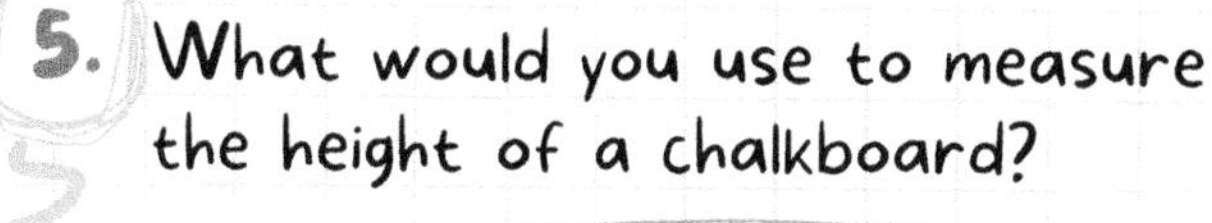

....

2. Show 9:12 on the clock:

....

....

3. Measure your thumb. How long is it in centimeters?

....

....

4. 74 + 27 = [ ]

$$\begin{array}{r} 74 \\ +\ 27 \\ \hline \end{array}$$

....

....

A. 91 B. 101 C. 111 D. 121

5. What would you use to measure the height of a chalkboard?

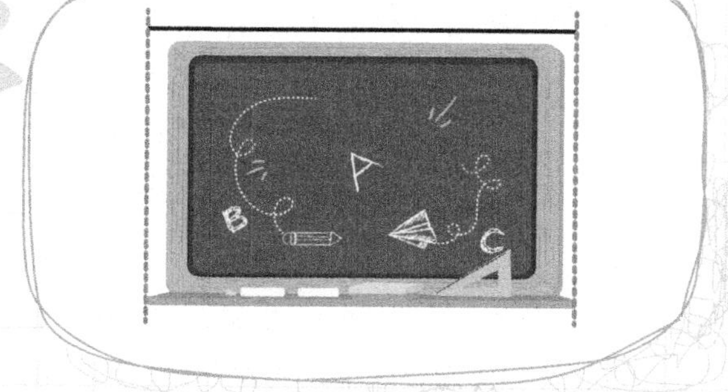

....

A. Feet B. Inches

6. How much is 3 quarters, 2 dimes and 7 pennies?

....

A. $1.02 B. $1.04 C. $1.06 D. $1.07

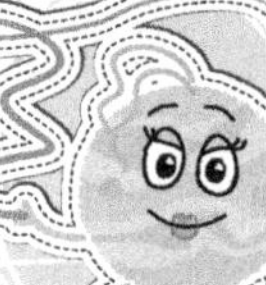

Week 12 Mixed Review

Topic 2 Numbers, operations, and time

1. 92 - 49 = ☐

$$\begin{array}{r} 92 \\ -\ 49 \\ \hline \end{array}$$

A. 53 B. 43 C. 33 D. 23

2. Show 2:58 on the clock.

3. Which number is not even?

A. 103 B. 106 C. 228 D. 230

4. Which number is not odd?

A. 853 B. 868 C. 811 D. 899

5. 17 - 4 = ☐

$$\begin{array}{r} 17 \\ -\ 4 \\ \hline \end{array}$$

A. 10 B. 12 C. 13 D. 15

6. 36 + 45 = ☐

$$\begin{array}{r} 36 \\ +\ 45 \\ \hline \end{array}$$

A. 71 B. 81 C. 91 D. 101

7. 86 - 31 = ☐

$$\begin{array}{r} 86 \\ -\ 31 \\ \hline \end{array}$$

A. 55 B. 45 C. 35 D. 25

8. 20 - 8 = ☐

$$\begin{array}{r} 20 \\ -\ 8 \\ \hline \end{array}$$

A. 11 B. 12 C. 13 D. 14

Read the following passage.

"Bunny!" Mama Rabbit called from downstairs. "It's time to get moving for your first day of school."

"Noooo!" wailed Bunny.

Mama and Papa Rabbit had taken Bunny to her new school last week to meet her teacher. But she was still feeling scared about being away from them for the first time. Her new teacher seemed nice enough, but it wasn't going to be the same as staying home with Mama. She pulled the covers tighter over her head and pretended to be asleep, as she heard Mama Rabbit's footsteps coming up the stairs.

"Bunny," Mama said softly, "I know you're feeling frightened today, but I promise everything is going to be just fine. You are going to love your new school. You will get to color and sing songs and meet new friends. It is going to be wonderful."

"But Mama, I'm going to miss you," Bunny replied. "What if I start feeling sad and cry in front of my new friends?"

"It's ok to feel sad, and it's ok to cry," Mama said. "Just remember, I will be there at 3 o'clock sharp to pick you up. Let's just give it a try. Can you do that for me?"

"OK," Bunny answered, still feeling very unsure.

She got up and began getting dressed anyways. When she arrived downstairs, she saw that Mama Rabbit had made her very favorite breakfast for her. She scurried to the table and gobbled up the french toast with carrots on top.

"Thanks Mama," Bunny said, "I think today WILL be ok. I'm ready to go!"

"All right," Mama Rabbit replied, smiling to herself as they walked out the door.

FITNESS PLANET
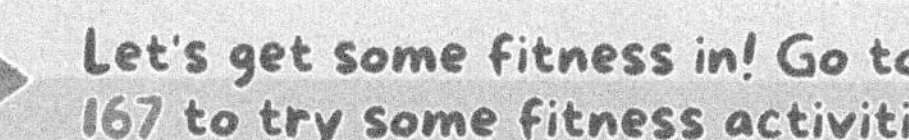
Let's get some fitness in! Go to page 167 to try some fitness activities.

Now answer the following questions about the passage.

1. What do you think the word "wailed" means in the passage?
   A. To sing
   B. To cry
   C. To smile

2. Why was Bunny feeling scared about her first day of school?
   A. She didn't like school.
   B. She didn't like her new teacher.
   C. She had never been away from Mama Rabbit before.

3. Why do you think Mama Rabbit made Bunny's favorite breakfast that morning?

   ....

   ....

4. What is Bunny afraid will happen at school?
   A. She will be sad and cry.
   B. The other students will make fun of her.
   C. She won't know anyone.

5. Complete the shades of meaning chart below for the given word "scared."

scared

6. What helps to change Bunny's mind about going to school?

   ....

   ....

7. Write a short paragraph about a time you felt scared about something.

   ....

   ....

Week 12 Mixed Review

Topic 3 Numbers, operations, and measurement

1. Which number is even? ....

A. 113   B. 123   C. 201   D. 164

2. What is 14 - 6?

....

A. 8   B. 10   C. 12   D. 14

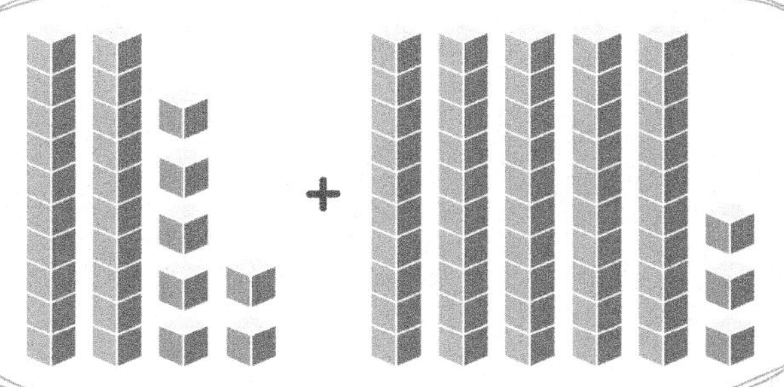

3. What is 27 + 53?

....

A. 80   B. 90   C. 100   D. 110

4. What would you use to measure the height of a bird's egg?

....

A. Meters   C. Inches
B. Centimeters   D. Feet

5. What is a reasonable length for a soda can?

....

A. 10 inches   C. 5 inches
B. 10 centimeters   D. 5 centimeters

6. What number on a clock would represent 45 minutes after the hour?

....

A. 9   B. 6   C. 3   D. 4

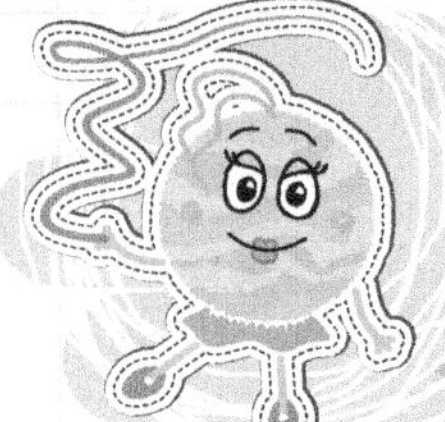

Week 12 Social Studies

Topic 3 Saving Money

Many people create and follow budgets in order to save money. They use these savings to pay bills, meet personal needs, and spend money for fun and entertainment. There are several different methods of saving money, and each has its own advantages and disadvantages.

Today, you are going to research a few of the most popular money saving methods and explain the pros and cons of each.

Money Saving Method	Advantages	Disadvantages
Piggy bank		
Savings account		
Investment account		

1. What do you think the best money saving method is? Why?

....

....

....

....

Grade 2-3

FITNESS PLANET

I ♡ FITNESS

I ♡ FITNESS

I ♡ FITNESS

FITNESS PLANET

Repeat these excersises 2 ROUNDS

exercises complex one

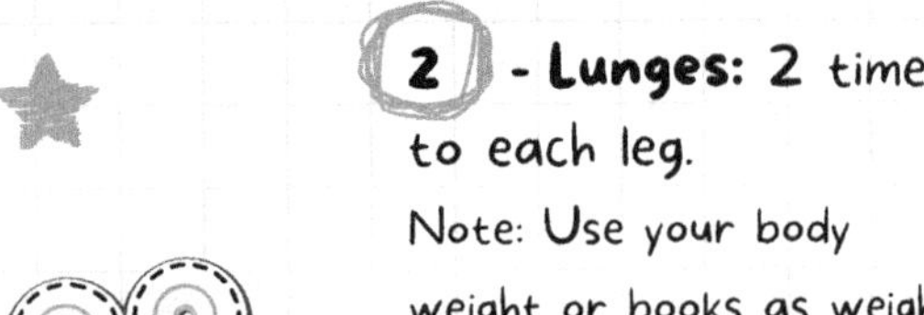

2 - **Lunges:** 2 times to each leg.
Note: Use your body weight or books as weight to do leg lunges.

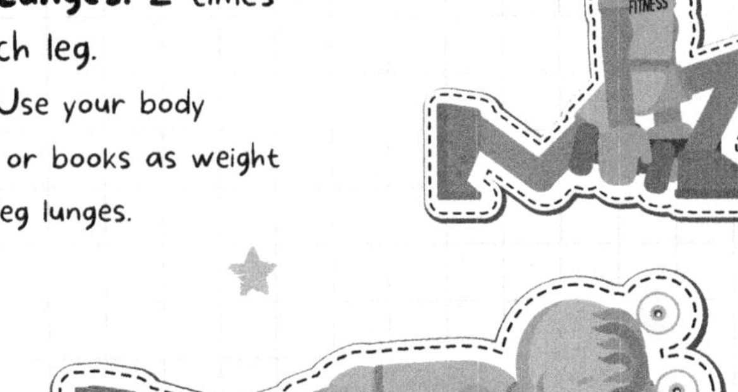

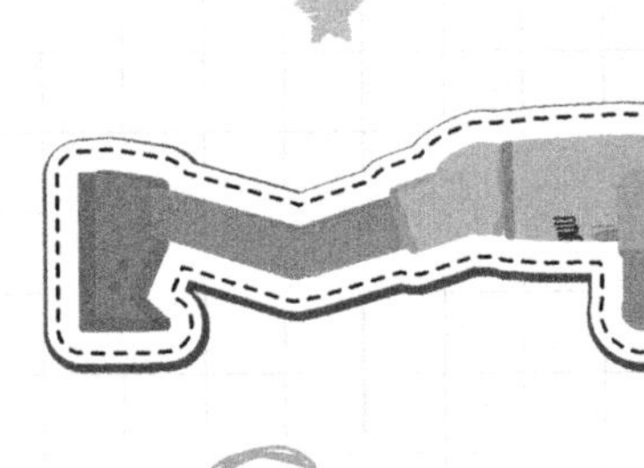

1 - **Abs:** 3 times

3 - **Plank:** 6 sec.

4 - **Run:** 50m
Note: Run **25** meters to one side and **25** meters back to the starting position.

Please be aware of your environment and be safe at all times. If you cannot do an exercise, just try your best.

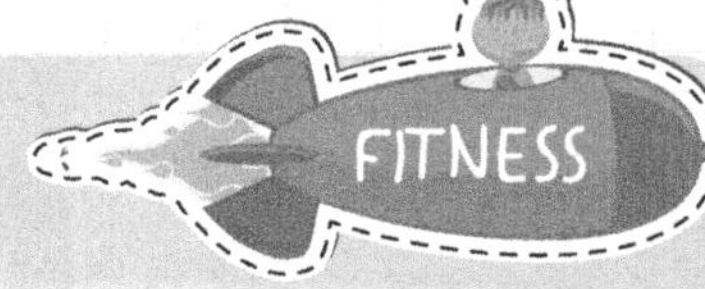

exercises complex two

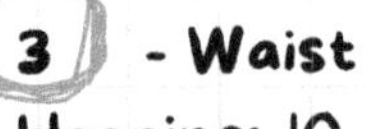

1 - **High Plank:** 6 sec.

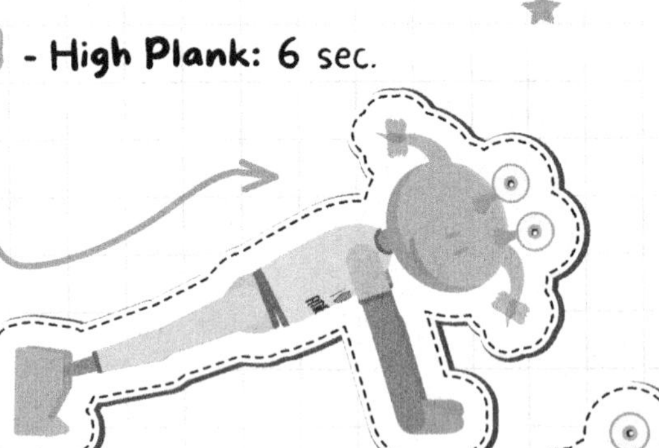

3 - **Waist Hooping:** 10 times.
Note: if you do not have a hoop, pretend you have an imaginary hoop and rotate your hips 10 times.

2 - **Chair:** 10 sec.
Note: sit on an imaginary chair, keep your back straight.

4 - **Abs:** 10 times

FITNESS PLANET

Repeat these excersises 2 ROUNDS

exercises complex three

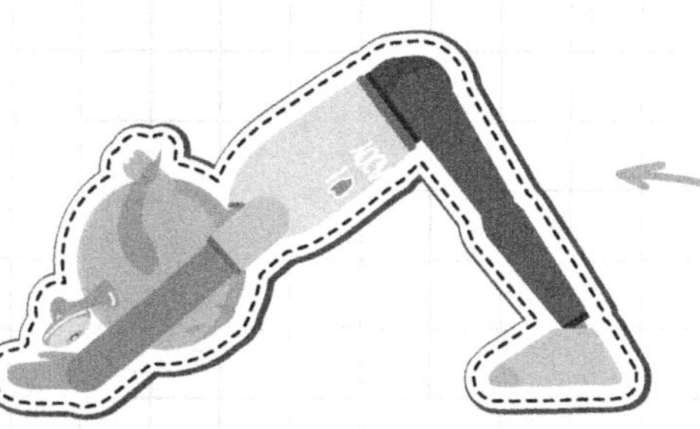

1 - **Down Dog:** 10 sec.

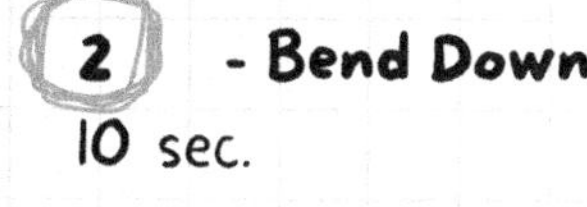

2 - **Bend Down:** 10 sec.

3 - **Chair:** 10 sec.

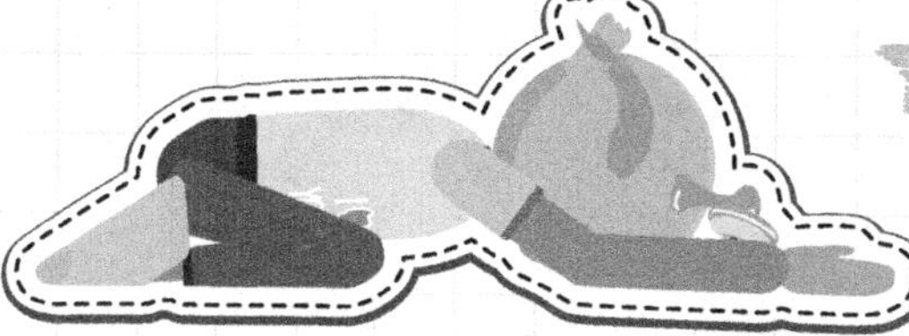

4 - **Child Pose:** 20 sec.

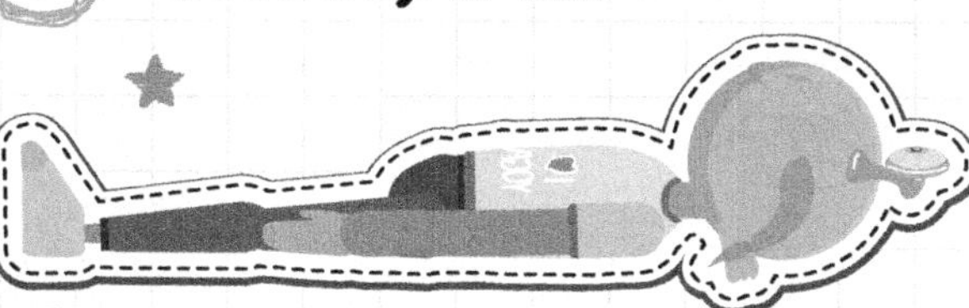

5 - **Shavasana:** as long as you can. Note: think of happy moments and relax your mind.

Please be aware of your environment and be safe at all times. If you cannot do an exercise, just try your best.

exercises complex four

1 - **Bend forward:** 10 times.
Note: try to touch your feet. Make sure to keep your back straight and if needed you can bend your knees.

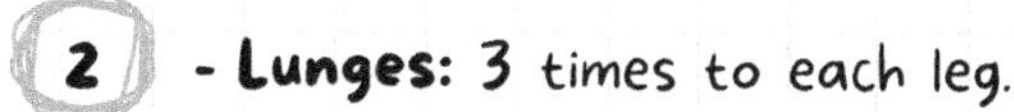

2 - **Lunges:** 3 times to each leg.
Note: Use your body weight or books as weight to do leg lunges.

3 - **Plank:** 6 sec.

4 - **Abs:** 10 times

Grade 2-3

ANSWERS SHEET

ANSWER SHEET

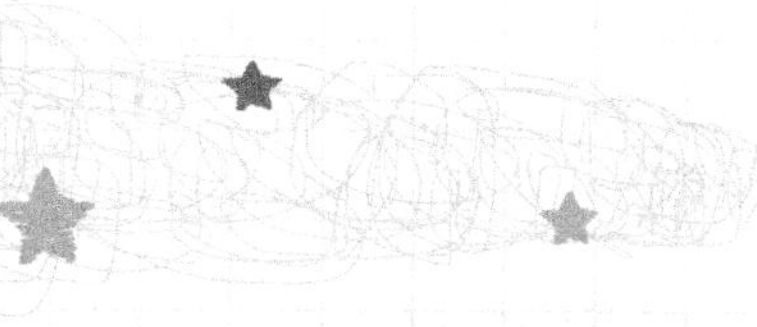

Week 1

Week 1 Operations and Algebraic Thinking

Topic 1 Mental Math: Add or Subtract to 20

Page 12	Page 13
1. B	1. A
2. A	2. A
3. C	3. D
4. D	4. D
5. A	5. B
	6. A

Topic 2 Addition and Subtraction Word Problems

Page 18	Page 19
1. A	1. A
2. B	2. B
3. B	3. D
4. C	4. D
5. C	5. B

Topic 3 Even or Odd

Page 22

1. Even	6. Odd	11. Even	16. Odd
2. Even	7. Odd	12. Even	17. Odd
3. Odd	8. Odd	13. Odd	18. Even
4. Even	9. Odd	14. Odd	19. Odd
5. Even	10. Even	15. Even	20. Even

Week 1 Language Review:

Topic 1 Parts of Speech

Page 14

1. Nouns: John, playground, swing
   Pronouns: he
   Verbs: ran, wanted
   Adjectives: blue
   Adverbs: quickly
2. Nouns: cat, ball
   Pronouns: she
   Verbs: stalked, pounced
   Adjectives: black, yarn
   Adverbs: mischievously

Page 15

1. D
2. Answers may vary.
3. Eating the rainbow means eating foods that are all different colors.
4. It is important to eat a variety of foods so that your body gets all the vitamins and nutrients it needs to stay healthy.
5. Protein, Dairy, Grains, Fruits, Vegetables, Oils

ANSWER SHEET

★Week 1 ★ Reading Passage

★ Topic 2 ★ Literature

★ Page 21 ★

1. D
2. Maxwell was feeling tired and hot in the middle of the story.
3. B
4. Carefully
5. Answers may vary.
6. Maxwell was so hot and tired from walking on a hot day that he laid down in the shade and refused to get up. Charlie needed to come up with a way to get Maxwell to walk back home with them.

★Week 1 ★ Science

★ Topic 1 ★ Investigating Simple Machines

★ Page 16-17 ★

1. The bean jumps higher with the marker than it does the pencil.
2. Yes, moving the fulcrum (pencil) toward the edge of the popsicle stick causes the bean to jump higher than it does when the fulcrum is in the middle.
3. The bean jumps the highest when the marker is used as the fulcrum and is placed toward the end of the popsicle stick.

★Week 1 ★ Exploring Communities

★ Topic 1 ★ Exploring Communities

★ Page 23 ★

1. Answers may vary.

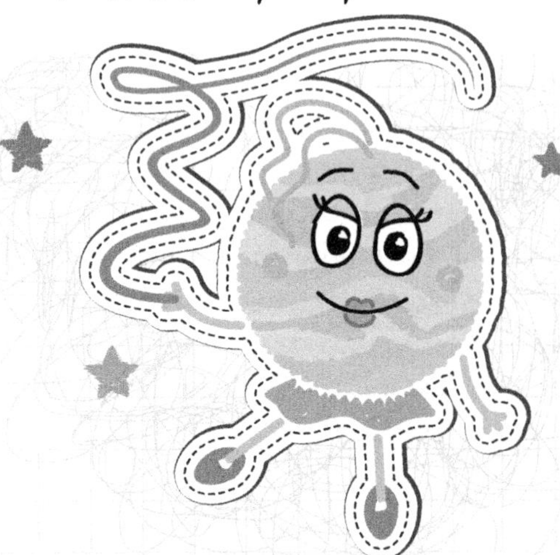

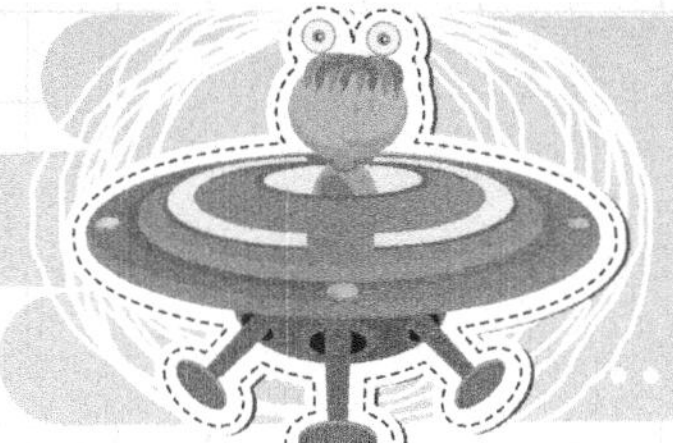

ANSWER SHEET

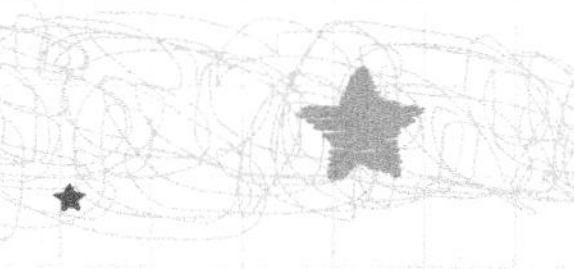

Week 2

Week 2 ★ Operations and Algebraic Thinking

Topic 1 ★ Building Arrays

Page 25

1. C
2. A
3. C
4. D

Page 26

1. A
2. A
3. A
4. B

Topic 2 ★ Representing Arrays with Equations

Page 31

1. 3 + 3 + 3 + 3 + 3 OR 5 + 5 + 5
2. 4 + 4 OR 2 + 2 + 2 + 2
3. 6 + 6 + 6 OR 3 + 3 + 3 + 3 + 3 + 3
4. 1 + 1 + 1 + 1 + 1
5. 3 + 3 + 3

Page 32

1. 2 + 2 + 2 + 2 + 2 OR 5 + 5
2. 9 + 9 + 9 OR
   3 + 3 + 3 + 3 + 3 +3 +3 + 3 + 3
3. 4 + 4 + 4 + 4
4. 7 + 7 OR 2+2+2+2+2+2+2
5. 4 + 4 + 4 + 4 + 4 + 4+ 4 + 4 OR
   7 + 7 + 7 + 7

Topic 3 ★ Using Arrays for Addition

Page 35

1. Students should draw shapes in 3 groups of 4.
2. Students should draw shapes in 2 groups of 2.
3. Students should draw shapes in 4 groups of 3.
4. Students should draw shapes in 3 groups of 1.
5. Students should draw shapes in 2 groups of 5.
6. Students should draw shapes in 6 groups of 2.

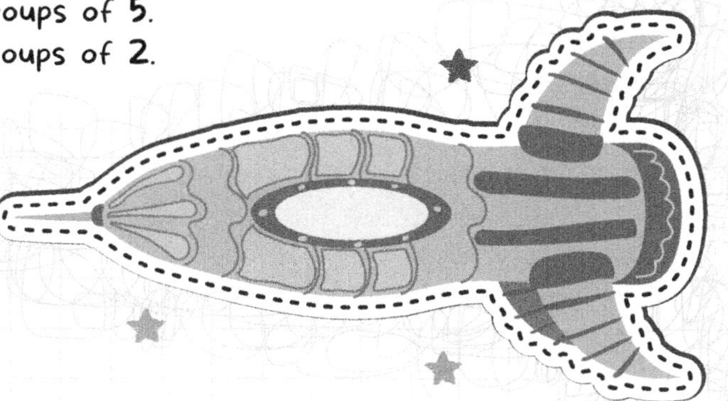

ANSWER SHEET

★Week 2 ★ Language Review

★ Topic 1 ★ Prefixes and Suffixes

★ Page 27 ★

1. Redo, do again
2. Preview, view before
3. Uncovered, not covered
4. Misheard, heard wrong
5. Discover, opposite of cover
6. Answers may vary.
7. Answers may vary.
8. A

★ Page 28 ★

1. Colorless, without color
2. Quickly, done in a quick manner
3. Doable, able to be done
4. Careful, full of care
5. Speaker, one who speaks
6. Answers may vary.
7. Answers may vary.
8. C

★Week 2 ★ Reading Passage

★ Topic 2 ★ Literature

★ Page 33 ★

1. Treat others the way you want to be treated, be kind to others, etc.
2. Maybe Shark really was jealous of Narwhal too.

★Week 2 ★ Reading Passage

★ Topic 3 ★ Informational Text

★ Page 34 ★

1. Table of contents, bold text, heading, glossary
2. Chapter 1
3. Chapter 3
4. What a koala looks like, what color fur koalas have, information about a koala's pouch, etc.
5. They are important keywords, or vocabulary words. They can probably be found in the glossary.
6. Answers may vary.
7. Answer may vary.

ANSWER SHEET

Week 2 ★ Science

Topic 1 ★ Investigating Rocks and Minerals

Page 29-30

1. Answers will vary.
2. Answers will vary.
3. When acids (vinegar) interact with calcium carbonate (limestone), carbon dioxide is released. When carbon dioxide is released in a liquid, bubbles form.

Week 2 ★ Exploring Communities

Topic 1 ★ Immigration

Page 36

1. 5000-10000 people
2. Many U.S. citizens can trace their heritage back to an ancestor that passed through Ellis Island; Ellis Island is how our country came to be a nation of immigrants.
3. It houses the Statue of Liberty.
4. Answers may vary.
5. Answers may vary.
6. Answers may vary.

Week 3

Week 3 ★ Operations and Algebraic Thinking

Topic 1 ★ Place value: Hundreds, Tens and Ones

Page 38

1. B
2. B
3. A
4. C
5. B
6. C
7. C

Page 39

1. B
2. B
3. A
4. C
5. C
6. B

Topic 2 ★ Counting and Skip Counting

Page 44

1. D
2. A
3. B
4. B
5. A
6. B
7. C
8. A

ANSWER SHEET

★ **Topic 3 ★ Reading and Writing Numbers to 1000**

★ **Page 45** ★

1. Forty-two
2. 800+20+1
3. 36
4. Two hundred thirty-one
5. 50+9
6. 486
7. 975
8. Three hundred seven
9. 600+70+4

★ **Page 48** ★

1. 642
2. Ninety-four
3. 400+7
4. 800+20+5
5. Fifty-one
6. 21
7. Seven hundred thirty-nine
8. 500+80+4
9. 83

★Week 3 ★ Language Review

★ **Topic 1 ★ Plural Nouns**

★ **Page 40** ★

1. Colors
2. Babies
3. Flashes
4. Knives
5. Vegetables
6. Fish
7. Women
8. Answers may vary.

★Week 3 ★ Reading Passage

★ **Topic 2 ★ Literature**

★ **Page 46** ★

1. Piece of cake, chewing him out
2. Literal text
3. He knew he hadn't done well on the test.
4. The author uses figurative language to make the text sound more dramatic.
5. Answers will vary.
6. Answers will vary.
7. Answers will vary.
8. Good luck
9. Be quiet
10. Answers will vary.

ANSWER SHEET

★Week 3 ★ Reading Passage

★ Topic 3 ★ Informational Text

★ Page 47 ★

1. Park, home, houses
2. Leaves are changing, leaves crunching, dusk, sun sinks lower, days are getting shorter, autumn
3. The breeze is chilly.
4. The leaves are falling from the trees.
5. Answers will vary.

★Week 3 ★ Science

★ Topic 1 ★ Exploring Fossils

★ Page 42-43 ★

1. Answers will vary.
2. Answers will vary.
3. The three slices of bread are flattened and stuck together.
4. The slices of bread are like the layers of sediment that harden into rock. Once the layers build up, you can't get them apart until erosion wears them down and uncovers the "fossil" inside.

★Week 3 ★ Exploring Communities

★ Topic 1 ★ Levels of Government

★ Page 49 ★

1. Answers may vary.
2. Answers may vary.
3. It depends upon the type of road (local road, state highway or interstate).

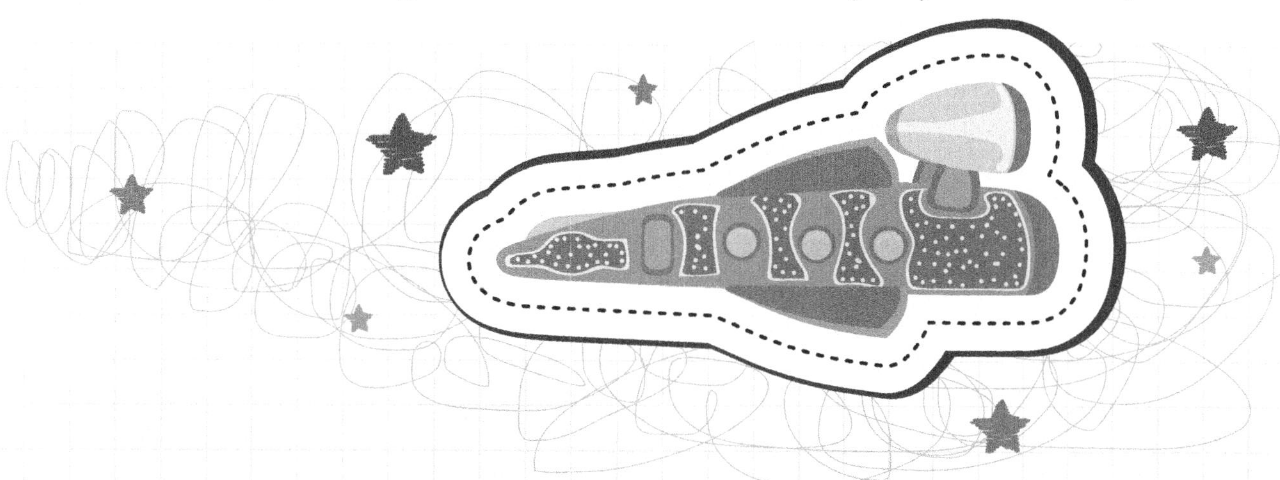

ANSWER SHEET

Week 4

Week 4 Number and Operations in Base Ten

Topic 1 Comparing Numbers

Page 51

1. D
2. A
3. B
4. C
5. C
6. A
7. D
8. B

Page 52

1. <
2. >
3. <
4. >
5. <
6. >
7. >
8. >
9. <
10. <

Topic 2 Adding Two Digit Numbers

Page 57

1. B
2. C
3. A
4. D
5. A
6. B
7. C

Topic 3 Adding up to 1000

Page 58

1. C
2. B
3. A
4. A
5. D
6. C
7. D
8. A

Page 61

1. C
2. B
3. A
4. A
5. D
6. B
7. C

ANSWER SHEET

Week 4 ★ Writing Review

★ Topic 1 ★ Opinion Writing

★ Page 53-54 ★

1. Answers will vary.

Week 4 ★ Reading Passage

★ Topic 2 ★ Literature

★ Page 59-60 ★

1. Mom
2. First person
3. Mom is telling the story and is also a character in the story.
4. Third person omniscient

Week 4 ★ Science

★ Topic 1 ★ Investigating Weather Patterns

★ Page 55-56 ★

1. Answers will vary.

Week 4 ★ Social Studies

★ Topic 1 ★ Government

★ Page 62 ★

1. A, C, D
2. B, C
3. Answers may vary.
4. Answers may vary.
5. Books, food, medicine
6. Fire and police protection, trash removal, road maintenance

ANSWER SHEET

Week 5

Week 5 Number and Operations in Base Ten

Topic 1 Mental Math: Adding and Subtracting by 10

Page 64

1. B
2. A
3. D
4. A
5. B

Page 65

1. B
2. A
3. A
4. D

Topic 2 Mental Math: Adding and Subtracting by 100

Page 70

1. C
2. B
3. D
4. B
5. A

Page 71

1. A
2. C
3. B
4. C
5. D

Topic 3 Mental Math: Adding and Subtracting by 10 and 100

Page 74

1. A
2. C
3. C
4. B
5. A

Week 5 Writing Review

Topic 1 Temporal Words

Page 66

Paragraph: Last week, first, suddenly, after, next, finally
Sentences: answers will vary

ANSWER SHEET

★ **Topic 1** ★ **Temporal Words**

★ **Page 67** ★

1. Dinosaurs, science, theater
2. D
3. A pretend thunderstorm
4. Fish and turtles
5. He loves fossils.
6. He built a boat.
7. B
8. The lights dimmed and he heard sounds like thunder.

★Week 5 ★ Reading Passage

★ **Topic 2** ★ **Informational Text**

★ **Page 72-73** ★

1. First person
2. I, we - the narrator is a character in the story
3. Third person - a narrator outside of the story
4. The child telling the story is actually a character in the story, and you are not, as the reader.
5. Today, after, as we were leaving
6. Answers may vary
7. Cute little animals, sloths are my new favorite animal

★ **Topic 2** ★ **Informational Text**

★ **Page 73** ★

The passage should be told by a narrator who knows the thoughts and feelings of one character - the young child in the story.

★Week 5 ★ Science

★ **Topic 1** ★ **Investigating Weather Hazards**

★ **Page 68-69** ★

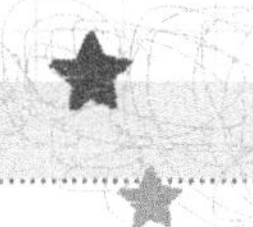

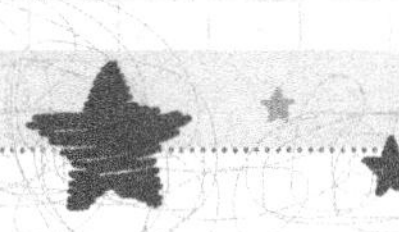

1. Midwest
2. Southeast, Mid-Atlantic
3. Midwest, Mid-Atlantic, Northeast
4. Typhoon, severe thunderstorms, high winds, forest fires
5. Hurricanes, severe thunderstorms, high winds, blizzards, ice storms

ANSWER SHEET

Week 5 ★ Social Studies

Topic 1 ★ Government Leaders

Page 75

Answers will vary.

Week 6

Week 6 ★ Measurement and Data

Topic 1 ★ Selecting and Using Appropriate Measurement Tools

Page 77

1. A
2. B
3. A
4. C
5. A

Page 78

1. A
2. A
3. C
4. B

Topic 2 ★ Estimate lengths

Page 83

1. A
2. B
3. A
4. A

Week 6 ★ Measuring an Object

Topic 3 ★ Measure each line using centimeters

Page 84

1. 6.5 cm
2. 3 cm
3. 1.5 cm
4. 5 cm
5. 4 cm
6. 2.5 cm
7. 4.5 cm
8. 2 cm

ANSWER SHEET

★ **Topic 3** ★ **Measure each line using inches**

★ **Page 87** ★

1. $4\frac{1}{4}$ in
2. $1\frac{1}{2}$ in
3. 2 in
4. $3\frac{1}{4}$ in
5. 5 in
6. $2\frac{1}{2}$ in
7. 1 in

★Week 6 ★ Writing Review

★ **Topic 1** ★ **Informative/Explanatory Writing**

★ **Page 79** ★

Answers will vary.

★Week 6 ★ Language Review

★ **Topic 2** ★ **Sentence Types**

★ **Page 80** ★

1. C
2. A
3. B
4. A
5. C

★Week 6 ★ Reading Passage

★ **Topic 3** ★ **Literature**

★ **Page 85-86** ★

1. Scared
2. Georgia was feeling scared as she lay in her bed, trying hard to fall asleep.
3. Relieved, not scared anymore
4. Scared, trying to fall asleep, pulled the covers up, squeezed her eyes closed, really wanted mom, about to scream, thank goodness, relaxed
5. A thunderstorm, scary sounds, shadows
6. She might continue to be scared or start screaming for mom.

ANSWER SHEET

Week 6 Science

Topic 1 Plant and Animal Traits

Page 82

1. Answers will vary.
2. Answers will vary.
3. The plants or flowers are of the same species, meaning they have similar, inherited traits.

Week 6 Social Studies

Topic 1 Geography Exploration

Page 88

Answers will vary.

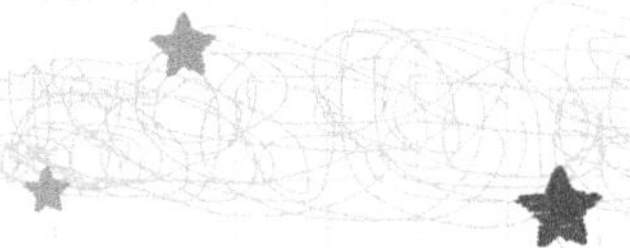

Week 7

Week 7 Measurement and Data

Topic 1 Looking at Length

Page 90

1. Mouse
2. Pen
3. Traffic light
4. Adult's shoe

Page 91

1. A
2. B
3. B
4. A
5. D

Topic 2 Whole Numbers on a Number Line

Page 96

1. A
2. C
3. D
4. B
5. A

ANSWER SHEET

★ Topic 3 ★ Telling Time

★ Page 97 ★

1. B
2. D
3. D
4. B
5. C
6. 12:02
7. 7:51

★ Page 100 ★

1. 4:48
2. 12:05
3. 10:17
4. 8:42
5. 3:33
6. Answers will vary
7. Answers will vary
8. Answers will vary

★ Week 7 ★ Language Review

★ Topic 1 ★ Spelling

★ Page 92 ★

1. D
2. A
3. C
4. A
5. B

★ Topic 1 ★ Spelling

★ Page 93 ★

1. Sitting
2. Happiness
3. Childhood
4. Shredded
5. Knives
6. Smiled
7. Beautiful

1. Babies
2. Leaves
3. Bushes
4. Teeth
5. Fish
6. Ladies
7. Children
8. Geese

ANSWER SHEET

Week 7 Reading Passage

Topic 2 Literature

Page 98-99

Chart: excited, having fun, looking forward to competing, nervous, in pain, embarrassed, comforted, thankful for her mom

1. She had been practicing the monkey bars.
2. If they were faster than the others on the monkey bars, the kids booed. If they were slower than the others, the kids cheered.
3. It means picked her up.
4. D
5. She falls off the monkey bars.
6. C
7. Her friend challenges her to a race.
8. All her friends are watching her.

Week 7 Science

Topic 1 The Basic Needs of Plants

Page 94-95

Answers will vary.

Week 7 Social Studies

Topic 1 Geography - Spatial Terms

Page 101

1. n/a
2. n/a
3. East
4. South
5. North America
6. Southern Hemisphere
7. Asia
8. Northern Hemisphere

ANSWER SHEET

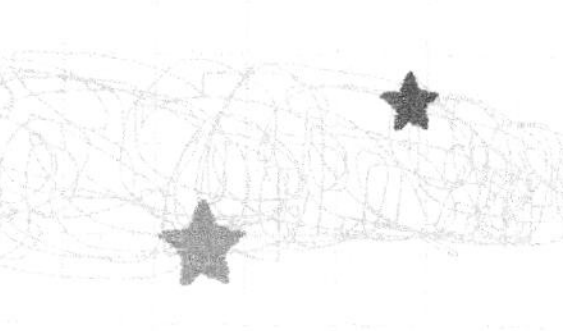

Week 8

Week 8 ★ Measurement and Data

Topic 1 ★ Word Problems Involving Money

Page 103

1. B
2. C
3. A
4. B
5. C

Page 104

1. Answers may vary. Student responses should add up to the correct total.
2. Answers may vary. Student responses should add up to the correct total.
3. Answers may vary. Student responses should add up to the correct total.
4. Answers may vary. Student responses should add up to the correct total.
5. Answers may vary. Student responses should add up to the correct total.
6. Answers may vary. Student responses should add up to the correct total.
7. Answers may vary. Student responses should add up to the correct total.
8. Answers may vary. Student responses should add up to the correct total.

Topic 2 ★ Generating Measurement Data

Page 109

1. Students should write down an object and a measurement that makes sense.
2. Students should write down an object and a measurement that makes sense.
3. Students should write down an object and a measurement that makes sense.
4. Students should write down an object and a measurement that makes sense.
5. Students should write down an object and a measurement that makes sense.
6. Students should write down an object and a measurement that makes sense.
7. Students should write down an object and a measurement that makes sense.
8. Students should write down an object and a measurement that makes sense.
9. Students should write down an object and a measurement that makes sense.

ANSWER SHEET

★ **Topic 3** ★ **Representing Data in Line, Picture and Bar Graphs**

★ **Page 110** ★

1. D
2. A
3. A
4. C
5. B

★ **Page 113** ★

1. D
2. C
3. B
4. C
5. A
6. B

★Week 8 ★ Writing Review

★ **Topic 1** ★ **Narrative Writing**

★ **Page 105-106** ★

Answers will vary.

★Week 8 ★ Language Review

★ **Topic 2** ★ **Commas in Addresses**

★ **Page 111**★

1. C
2. B
3. 2708 Main Street, Elmhurst, Illinois 60807

★Week 8 ★ Reading Passage

★ **Topic 2** ★ **Informational Text**

★ **Page 112**★

1. C
2. The paragraph describes the steps in a process in sequential order.
3. Transition words
4. First, next, then, after that, lastly
5. Second, third, finally, in the beginning
6. They help the reader understand the order of steps or events in the passage.

ANSWER SHEET

Week 8 Science

Topic 1 A Closer Look at Animals

Page 107-108

1. Weaker animals can live and travel in groups to help protect themselves from predators.
2. Fish, birds
3. Some animals hunt in groups. They work together to kill their prey.
4. Wolves, dolphins, lions
5. Some animals live in groups and cooperate to care for the young in the group.
6. Monkeys, rats, whales
7. Some animals live in groups and cooperate to build shelter. This makes it easier, as many animals are working together.
8. Beavers

Week 8 Social Studies

Topic 1 Regions

Page 1014

Answers will vary.

Week 9

Week 9 Geometry

Topic 1 Attributes of Shapes

Page 116

1. A
2. C
3. B
4. D
5. D
6. D
7. B

Page 117

1. Student should create a shape with three sides.
2. Student should create a shape with four sides.
3. Student should create a shape with five sides.
4. Student should create a shape with six sides.
5. Student should create a three dimensional shape with six sides.
6. Student should create a shape with three sides.

ANSWER SHEET

★ **Topic 2** ★ **Dividing Rectangles**

★ **Page 122** ★

1. Students should divide this rectangle into equal parts.
2. B
3. A
4. D
5. C
6. Students should divide this rectangle into equal parts.

★ **Topic 3** ★ **Dividing circles**

★ **Page 123** ★

1. Students should divide this circle into 2 equal parts.
2. A
3. Students should divide this circle into 4 equal parts.
4. Students should divide this circle into 6 equal parts and shade in 2.
5. Students should divide this circle into 4 equal parts and shade in 2.
6. Students should divide this circle into 3 equal parts.
7. Students should divide this circle into 8 equal parts and shade in 4.

★ **Topic 4** ★ **Connecting shapes to fractions.**

★ **Page 126** ★

1. B
2. B
3. C
4. C
5. D
6. A

★Week 9 ★ Language Reviewy

★ **Topic 1** ★ **Verb Tenses**

★ **Page 118-119** ★

1. A
2. B
3. C
4. B
5. Cried
6. Ran
7. Answers will vary.
8. Answers will vary.
9. Answers will vary.
10. Future tense
11. Juan played in a soccer game last Saturday. Juan is playing in a soccer game.
12. Raul ran down the street to catch up with his friends.

ANSWER SHEET

Week 9 ★ Reading Passage

Topic 2 ★ Literature

Page 124-125

1. The princess, the king, Snowflake
2. B
3. C
4. The princess is getting ready for the ball, but her tiara is missing.
5. Her dog, Snowflake
6. She remembers seeing Snowflake acting strangely in the rose garden the week before.
7. Naughty, mischievious
8. Answers will vary.

Week 9 ★ Science

Topic 1 ★ Properties of Sound

Page 120-121

Answers will vary.

Week 9 ★ Social Studies

Topic 1 ★ Climate Regions

Page 127

1. Northeast: bitterly cold winters, humid summers, blizzards, occasional hurricanes
2. Southeast: humid, very hot summers, warm winters, hurricanes
3. West: cool, wet winters and dry, hot summers
4. Southwest: warm winters and dry, hot summers
5. Midwest: humid, hot summers and cold winters, tornadoes

Week 10

Week 10 ★ More Calculations

Topic 1 ★ Addition and subtraction review

Page 129

1. B
2. D
3. C
4. C
5. B
6. D

Page 130

1. C
2. B
3. D
4. B
5. C
6. A

ANSWER SHEET

★ **Topic 2** ★ **Addition and subtraction review**

★ **Page 135** ★

1. B
2. A
3. D
4. C
5. A
6. D

★ **Page 136** ★

1. C
2. B
3. C
4. A
5. C
6. A

★ **Topic 3** ★ **Addition and subtraction review**

★ **Page 139** ★

1. B
2. B
3. C

★ Week 10 ★ Language Review

★ **Topic 1** ★ **Quotation Marks**

★ **Page 131-132** ★

1. "The weather is turning colder," Montrell said.
2. "Yikes! I'm scared of spiders," she shouted.
3. "Is it lunchtime yet?" he asked the teacher.
4. "I am so excited about this weekend. My mom said we could go to the store to buy a new dress for the party. I hope I find a pink one," Joelle told Michelle.
5. "No!" screamed Marco.
6. "Did you bring a coat?" the teacher asked.
7. "This field trip," Hugo exclaimed, "is going to be the best one yet!"
8. "What should I do next?" she asked her dad.
9. "Let's go!" mom shouted up the stairs.
10. "On Thursday," mom stated, "we will go to the store."

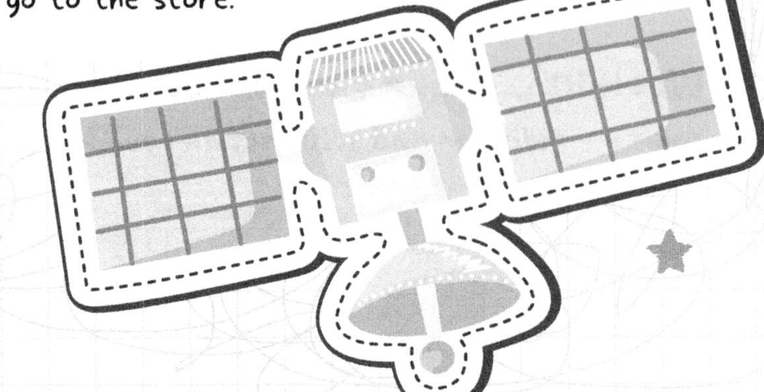

ANSWER SHEET

Week 10 Reading Passage

Topic 1 Informational Text

Page 137-138

1. Passage 1 - Plants are an important part of the ecosystem.
   Passage 2 - Growing plants is not hard.
2. Passage 1 - Plants provide many benefits to humans and animals. Without plants, humans and animals would not have everything they need to survive.
   Passage 2 - Plants need water, air, sunlight and nutrients. You should water your seed every day.
3. Both passages are about plants and discuss the needs of plants and animals in an ecosystem.
4. The first passage discusses the benefits plants provide to other animals in the ecosystem while the second passage discusses what plants need in order to survive.
5. Plants are important.
6. Some small insects and animals may live under or in the leaves. For example, birds often build their nests in the leaves of trees.
7. Plant a tree, water a garden, fertilize the soil

Week 10 Sciencee

Topic 1 How Sound Travels

Page 133-134

1. You hear a sound.
2. The ruler hitting the spoon creates vibrations which move up the yarn and can be heard by the person holding the yarn up to their ears.
3. The sound is higher pitched or deeper, depending upon the size of the spoon.
4. No, only the person holding the string up to their ears can hear the noise because the sound waves are traveling up the yarn and not through the air.

Week 10 Social Studies

Topic 1 Exploring Economics in Your Community

Page 140

All answers will vary.

ANSWER SHEET

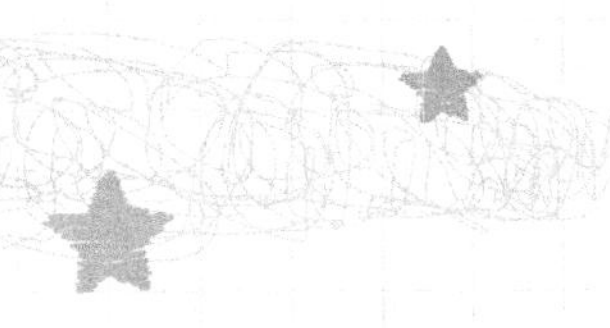

Week 11

Week 11 ★ Making Calculations Mixed Review

Topic 1 ★ Odd and even numbers

Page 142

1. Odd
2. Odd
3. Even
4. Even
5. Even
6. Even
7. Even
8. Even
9. Odd
10. Even

Page 143

1. 19
2. 16
3. 19
4. 16
5. 17
6. 17
7. 10
8. 5
9. 4
10. 12

Topic 2 ★ Two-digit addition and subtraction review

Page 148

1. B
2. C
3. A
4. B
5. C
6. B
7. B

Page 149

1. 20
2. 500
3. 160
4. 730
5. 80
6. 70
7. 405
8. 310
9. 50
10. 975

Topic 3 ★ Addition and subtraction under 1000

Page 152

1. B
2. C
3. B
4. B
5. C
6. A
7. A

ANSWER SHEET

★Week 11 ★ Language Review

★ Topic 1 ★ Context Clues

★ Page 144-145 ★

1. Worth a lot of money
2. Nice, kind
3. A break

1. A
2. D
3. A

★Week 11 ★ Reading Passage

★ Topic 2 ★ Literature

★ Page 150 ★

1. B
2. A
3. He tries to teach them so that they can improve.

★Week 11 ★ Sciencee

★ Topic 1 ★ Can Sound Travel Through Liquids?

★ Page 146-147 ★

1. You can hear the sounds of the spoons tapping against one another.
2. Sound waves are able to travel through water.
3. Sound travels faster through water because the particles in water are much closer together than they are in the air.

★Week 11 ★ Social Studies

★ Topic 1 ★ Exploring Economics in Your Community

★ Page 153 ★

Costs: Lost sleep when training a puppy, obedience classes, food costs, grooming costs, dog sitting fees when traveling, having to walk a dog in cold or rainy weather

Benefits: Offers protection, additional exercise from walking a dog, stress reliever, brings the family together, teaches responsibility

1. Answers will vary.

ANSWER SHEET

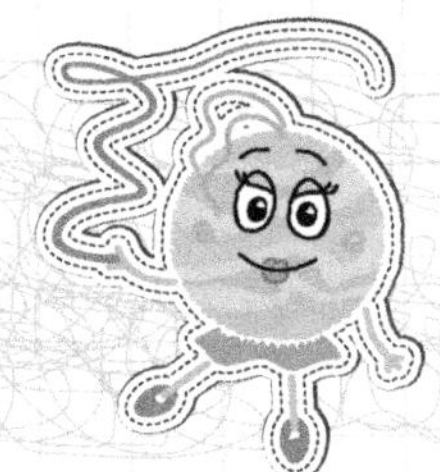

Week 12

Week 12 Mixed Review

Topic 1 Numbers, operations, measurement and time

Page 155

1. C
2. A
3. Students should label an object and measure it.
4. C
5. B
6. A
7. Students should draw the hour hand pointing to the one and the minute hand after the 7.

Page 156

1. 2:00 PM
2. B
3. C
4. B
5. C
6. A

Topic 2 Numbers, operations, and time

Page 161

1. A lamp
2. The hour hand should point to the 9 and the minute hand should point after the 2.
3. Students should write a realistic measurement.
4. B
5. Feet
6. D

Page 162

1. B
2. Students should place the hour hand at the 2 and the minute hand after the 11.
3. A
4. B
5. C
6. B
7. A
8. B

Topic 3 Numbers, operations, and measurement

Page 165

1. D
2. A
3. A
4. B
5. D
6. A

ANSWER SHEET

Week 12 Language Review

Topic 1 Shades of Meaning

Page 157

1. Happy, thrilled, ecstatic, glad, joyful, delighted
2. Said, whispered, declared, stated, questioned, asked, shouted
3. Walked, skipped, sauntered, strolled, ran, ambled

Page 158

Answers will vary.

Week 12 Reading Passage

Topic 2 Literature

Page 163-164

1. B
2. C
3. To help her feel better about her first day of school.
4. A
5. Scared, terrified, frightened, afraid, anxious, nervous
6. Her mom made her favorite breakfast.
7. Answers will vary.

Week 12 Science

Topic 1 How Does Sound Travel Through Solids?

Page 159-60

1. No, because the sound doesn't travel fast enough through the air to reach the other person.
2. Yes, because the sound travels through the solid (the string) and allows the other person to hear it in the container.
3. Sound travels fastest through solids because the particles are closest together.

Week 12 Social Studies

Topic 1 Saving Money

Page 166

All answers will vary.

Made in the USA
Columbia, SC
21 June 2020